An Introduction to Complex Analysis and Geometry

The Sally SERIES

Pure and Applied
UNDERGRADUATE TEXTS · 12

An Introduction to Complex Analysis and Geometry

John P. D'Angelo

American Mathematical Society
Providence, Rhode Island

2010 *Mathematics Subject Classification.* Primary 30–00, 30–01, 51–01, 51M99, 26–01, 40–01.

For additional information and updates on this book, visit
www.ams.org/bookpages/amstext-12

Library of Congress Cataloging-in-Publication Data

D'Angelo, John P.
    An introduction to complex analysis and geometry / John P. D'Angelo.
      p. cm. — (Pure and applied undergraduate texts ; v. 12)
    Includes bibliographical references and index.
    ISBN 978-0-8218-5274-3 (alk. paper)
    1. Functions of complex variables. 2. Geometry, Algebraic. 3. Sequences (Mathematics)
I. Title.

QA331.7.D356   2011
515′.9—dc22

                                                      2010029859

Contents

Preface

This book developed from a course given in the Campus Honors Program at the University of Illinois Urbana-Champaign in the fall semester of 2008. The aims of the course were to introduce bright students, most of whom were freshmen, to complex numbers in a friendly, elegant fashion and to develop reasoning skills belonging to the realm of elementary complex geometry. In the spring semester of 2010 I taught another version of the course, in which a draft of this book was available online. I therefore wish to acknowledge the Campus Honors Program at UIUC for allowing me to teach these courses and to thank the 27 students who participated in them.

Many elementary mathematics and physics problems seem to simplify magically when viewed from the perspective of complex analysis. My own research interests in functions of several complex variables and CR geometry have allowed me to witness this magic daily. I continue the preface by mentioning some of the specific topics discussed in the book and by indicating how they fit into this theme.

Every discussion of complex analysis must spend considerable time with power series expansions. We include enough basic analysis to study power series rigorously and to solidify the backgrounds of the typical students in the course. In some sense two specific power series dominate the subject: the geometric and exponential series.

The geometric series appears all throughout mathematics and physics and even in basic economics. The Cauchy integral formula provides a way of deriving from the geometric series the power series expansion of an arbitrary complex analytic function. Applications of the geometric series appear throughout the book.

The exponential series is of course also crucial. We define the exponential function via its power series, and we define the trigonometric functions by way of the exponential function. This approach reveals the striking connections between the functional equation $e^{z+w} = e^z e^w$ and the profusion of trigonometric identities.

Using the complex exponential function to simplify trigonometry is a compelling aspect of elementary complex analysis and geometry. Students in my courses seemed to appreciate this material to a great extent.

One of the most appealing combinations of the geometric series and the exponential series appears in Chapter 4. We combine them to derive a formula for the sums

$$\sum_{j=1}^{n} j^p,$$

in terms of Bernoulli numbers.

We briefly discuss ordinary and exponential generating functions, and we find the ordinary generating function for the Fibonacci numbers. We then derive Binet's formula for the n-th Fibonacci number and show that the ratio of successive Fibonacci numbers tends to the golden ratio $\frac{1+\sqrt{5}}{2}$.

Fairly early in the book (Chapter 3) we discuss hyperbolas, ellipses, and parabolas. Most students have seen this material in calculus or even earlier. In order to make the material more engaging, we describe these objects by way of Hermitian symmetric quadratic polynomials. This approach epitomizes our focus on complex numbers rather than on pairs of real numbers.

The geometry of the unit circle also allows us to determine the Pythagorean triples. We identify the Pythagorean triple (a, b, c) with the complex number $\frac{a}{c} + i\frac{b}{c}$; we then realize that a Pythagorean triple corresponds to a rational point (in the first quadrant) on the unit circle. After determining the usual rational parametrization of the unit circle, one can easily find all these triples. But one gains much more; for example, one discovers the so-called $\tan(\frac{\theta}{2})$ substitution from calculus. During the course several students followed up this idea and tracked down how the indefinite integral of the secant function arose in navigation.

This book is more formal than was the course itself. The list of approximately two hundred eighty exercises in the book is also considerably longer than the list of assigned exercises. These exercises (as well as the figures) are numbered by chapter, whereas items such as theorems, propositions, lemmas, examples, and definitions are numbered by section. Unless specified otherwise, a reference to a section, theorem, proposition, lemma, example, or definition is to the current chapter. The overall development in the book closely parallels that of the courses, although each time I omitted many of the harder topics. I feel cautiously optimistic that this book can be used for similar courses. Instructors will need to make their own decisions about which subjects can be omitted. I hope however that the book has a wider audience including anyone who has ever been curious about complex numbers and the striking role they play in modern mathematics and science.

Chapter 1 starts by considering various number systems and continues by describing, slowly and carefully, what it means to say that the real numbers are a complete ordered field. We give an interesting proof that there is no *rational* square root of 2, and we prove carefully (based on the completeness axiom) that positive real numbers have square roots. The chapter ends by giving several possible definitions of the field of complex numbers.

Chapter 2 develops the basic properties of complex numbers, with a special emphasis on the role of complex conjugation. The author's own research in complex analysis and geometry has often used *polarization*; this technique makes precise the sense in which we may treat z and $\bar{z}$ as independent variables. We will view complex analytic functions as those independent of $\bar{z}$. In this chapter we also include precise definitions about convergence of series and related elementary analysis. Some instructors will need to treat this material carefully, while others will wish to review it quickly. Section 5 treats uniform convergence and some readers will wish to postpone this material. The subsequent sections however return to the basics of complex geometry. We define the exponential function by its power series and the cosine and sine functions by way of the exponential function. We can and therefore do discuss logarithms and trigonometry in this chapter as well.

Chapter 3 focuses on geometric aspects of complex numbers. We analyze the zero-sets of quadratic equations from the point of view of complex rather than real variables. For us hyperbolas, parabolas, and ellipses are zero-sets of quadratic Hermitian symmetric polynomials. We also study linear fractional transformations and the Riemann sphere.

Chapter 4 considers power series in general; students and instructors will find that this material illuminates the treatment of series from calculus courses. The chapter includes a short discussion of generating functions, Binet's formula for the Fibonacci numbers, and the formula for sums of p-th powers mentioned above. We close Chapter 4 by giving a test for when a power series defines a rational function.

Chapter 5 begins by posing three possible definitions of complex analytic function. These definitions involve locally convergent power series, the Cauchy-Riemann equations, and the limit quotient version of complex differentiability. We postpone the proof that these three definitions determine the same class of functions until Chapter 6 after we have introduced integration. Chapter 5 focuses on the relationship between real and complex derivatives. We define the Cauchy-Riemann equations using the $\frac{\partial}{\partial \bar{z}}$ operator. Thus complex analytic functions are those functions *independent* of $\bar{z}$. This perspective has profoundly influenced research in complex analysis, especially in higher dimensions, for at least fifty years. We briefly consider harmonic functions and differential forms in Chapter 5; for some audiences there might be too little discussion about these topics. It would be nice to develop potential theory in detail and also to say more about closed and exact differential forms, but then perhaps too many readers would drown in deep water.

Chapter 6 treats the Cauchy theory of complex analytic functions in a simplified fashion. The main point there is to show that the three possible definitions of analytic function introduced in Chapter 5 all lead to the same class of functions. This material forms the basis for both the theory and application of complex analysis. In short, Chapter 5 considers derivatives and Chapter 6 considers integrals.

Chapter 7 offers many applications of the Cauchy theory to ordinary integrals. In order to show students how to apply complex analysis to things they have seen before, we evaluate many interesting real integrals using residues and contour integration. We also include sections on the Fourier transform on the Gamma function.

Chapter 8 introduces additional appealing topics such as the fundamental theorem of algebra (for which we give three proofs), winding numbers, Rouche's theorem, Pythagorean triples, conformal mappings, the quaternions, and (a brief mention of) complex analysis in higher dimensions. The section on conformal mappings
includes a brief discussion of non-Euclidean geometry. The section on quaternions
includes the observation that there are many quaternionic square roots of -1, and
hence it illuminates the earliest material used in defining $\mathbf{C}$. The final result proved
concerns polarization; it justifies treating z and $\overline{z}$ as independent variables, and
hence it also unifies much of the material in the book.

Our bibliography includes many excellent books on complex analysis in one
variable. One naturally asks how this book differs from those. The primary difference is that this book begins at a more elementary level. We start at the logical
beginning, by discussing the natural numbers, the rational numbers, and the real
numbers. We include detailed discussion of some truly basic things, such as the
existence of square roots of positive real numbers, the irrationality of $\sqrt{2}$, and
several different definitions of $\mathbf{C}$ itself. Hence most of the book can be read by
a smart freshman who has had some calculus, but not necessarily any real analysis. A second difference arises from the desire to engage an audience of bright
freshmen. I therefore include discussion, examples, and exercises on many topics
known to this audience via real variables, but which become more transparent using complex variables. My ninth grade mathematics class (more than forty years
ago) was tested on being able to write word-for-word the definitions of hyperbola,
ellipse, and parabola. Most current college freshmen know only vaguely what these
objects are, and I found myself reciting those definitions when I taught the course.
During class I also paused to carefully prove that .999... really equals 1. Hence
the book contains various basic topics, and as a result it enables spiral learning.
Several concepts are revisited with high multiplicity throughout the book. A third
difference from the other books arises from the inclusion of several unusual topics,
as described throughout this preface.

I hope, with some confidence, that the text conveys my deep appreciation for
complex analysis and geometry. I hope, but with more caution, that I have purged
all errors from it. Most of all I hope that many will enjoy reading it and solving
the exercises in it.

I began expanding the sketchy notes from the course into this book during
the spring 2009 semester, during which I was partially supported by the Kenneth
D. Schmidt Professorial Scholar award. I therefore wish to thank Dr. Kenneth
Schmidt and also the College of Arts and Sciences at UIUC for awarding me this
prize. I have received considerable research support from the NSF for my work
in complex analysis; in particular I acknowledge support from NSF grant DMS-
0753978. The students in the first version of the course survived without a text;
their enthusiasm and interest merit praise. Over the years many other students
have inspired me to think carefully how to present complex analysis and geometry
with elegance. Another positive influence on the evolution from sketchy notes to
this book was working through some of the material with Bill Heiles, Professor
of Piano at UIUC and one who appreciates the art of mathematics. Jing Zou,
computer science student at UIUC, prepared the figures in the book. Tom Forgacs,

who invited me to speak at California State University, Fresno on my experiences teaching this course, also made useful comments. My colleague Jeremy Tyson made many valuable suggestions on both the mathematics and the exposition. I asked several friends to look at the N-th draft for various large N. Bob Vanderbei, Rock Rodini, and Mike Bolt all made many useful comments which I have incorporated. I thank Sergei Gelfand and Ed Dunne of the American Mathematical Society for encouragement; Ed Dunne provided me marked-up versions of two drafts and shared with me, in a lengthy phone conversation, his insights on how to improve and complete the project. Cristin Zannella and Arlene O'Sean of the AMS oversaw the final editing and other finishing touches. Finally I thank my wife Annette and our four children for their love.

Preface for the student

I hope that this book reveals the beauty and usefulness of complex numbers to you. I want you to enjoy both reading it and solving the problems in it. Perhaps you will spot something in your own area of interest and benefit from applying complex numbers to it. Students in my classes have found applications of ideas from this book to physics, music, engineering, and linguistics. Several students have become interested in historical and philosophical aspects of complex numbers. I have not yet seen anyone get excited about the hysterical aspects of complex numbers.

At the very least you should see many places where complex numbers shed a new light on things you have learned before. One of my favorite examples is trig identities. I found them rather boring in high school and later I delighted in proving them more easily using the complex exponential function. I hope you have the same experience. A second example concerns certain definite integrals. The techniques of complex analysis allow for stunningly easy evaluations of many calculus integrals and seem to lie within the realm of science fiction.

This book is meant to be readable, but at the same time it is precise and rigorous. Sometimes mathematicians include details that others feel are unnecessary or obvious, but do not be alarmed. If you do many of the exercises and work through the examples, then you should learn plenty and enjoy doing it. I cannot stress enough two things I have learned from years of teaching mathematics. First, students make too few sketches. You should strive to merge geometric and algebraic reasoning. Second, definitions are your friends. If a theorem says something about a concept, then you should develop both an intuitive sense of the concept and the discipline to learn the precise definition. When asked to verify something on an exam, start by writing down the definition of that something. Often the definition suggests exactly what you should do!

Some sections and paragraphs introduce more sophisticated terminology than is necessary at the time, in order to prepare for later parts of the book and even for subsequent courses. I have tried to indicate all such places and to revisit the crucial ideas. In case you are struggling with any material in this book, remain calm. The magician will reveal his secrets in due time.

Chapter 1

From the Real Numbers
to the Complex Numbers

1. Introduction

Many problems throughout mathematics and physics illustrate an amazing principle: ideas expressed within the realm of real numbers find their most elegant expression through the unexpected intervention of complex numbers. Many of these delightful interventions arise in elementary, recreational mathematics. On the other hand most college students either never see complex numbers in action or they wait until the junior or senior year in college, at which time the sophisticated courses have little time for the elementary applications. Hence too few students witness the beauty and elegance of complex numbers. This book aims to present a variety of elegant applications of complex analysis and geometry in an accessible but precise fashion. We begin at the beginning, by recalling various number systems such as the integers $\mathbf{Z}$, the rational numbers $\mathbf{Q}$, and the real numbers $\mathbf{R}$, before even defining the complex numbers $\mathbf{C}$. We then provide three possible equivalent definitions. Throughout we strive for as much geometric reasoning as possible.

2. Number systems

The ancients were well aware of the so-called *natural numbers*, written $1, 2, 3, \ldots$. Mathematicians write $\mathbf{N}$ for the collection of natural numbers together with the usual operations of addition and multiplication. Partly because subtraction is not always possible, but also because negative numbers arise in many settings such as financial debts, it is natural to expand the natural number system to the larger system $\mathbf{Z}$ of integers. We assume that the reader has some understanding of the integers; the set $\mathbf{Z}$ is equipped with two distinguished members, written 1 and 0, and two operations, called addition (+) and multiplication ($*$), satisfying familiar laws. These operations make $\mathbf{Z}$ into what mathematicians call a *commutative ring with unit* 1. The integer 0 is special. We note that each n in $\mathbf{Z}$ has an additive inverse $-n$ such that

(1) $$n + (-n) = (-n) + n = 0.$$

Of course 0 is the only number whose additive inverse is itself.

Let a, b be given integers. As usual we write $a - b$ for the sum $a + (-b)$. Consider the equation $a + x = b$ for an unknown x. We learn to solve this equation at a young age; the idea is that subtraction is the inverse operation to addition. To solve $a + x = b$ for x, we first add $-a$ to both sides and use (1). We can then substitute b for $a + x$ to obtain the solution

$$x = 0 + x = (-a) + a + x = (-a) + b = b + (-a) = b - a.$$

This simple principle becomes a little more difficult when we work with multiplication. It is not always possible, for example, to divide a collection of n objects into two groups of equal size. In other words, the equation $2 * a = b$ does not have a solution in $\mathbf{Z}$ unless b is an even number. Within $\mathbf{Z}$, most integers (± 1 are the only exceptions) do not have multiplicative inverses.

To allow for division, we enlarge $\mathbf{Z}$ into the larger system $\mathbf{Q}$ of rational numbers. We think of elements of $\mathbf{Q}$ as fractions, but the definition of $\mathbf{Q}$ is a bit subtle. One reason for the subtlety is that we want $\frac{1}{2}$, $\frac{2}{4}$, and $\frac{50}{100}$ all to represent the same rational number, yet the expressions as fractions differ. Several approaches enable us to make this point precise. One way is to introduce the notion of equivalence class and then to define a rational number to be an equivalence class of pairs of integers. See [4] or [8] for this approach. A second way is to think of the rational number system as known to us; we then write elements of $\mathbf{Q}$ as letters, x, y, u, v, and so on, without worrying that each rational number can be written as a fraction in infinitely many ways. We will proceed in this second fashion. A third way appears in Exercise 1.2 below. Finally we emphasize that we cannot divide by 0. Surely the reader has seen alleged proofs that, for example, $1 = 2$, where the only error is a cleverly disguised division by 0.

▶ **Exercise 1.1.** Find an invalid argument that $1 = 2$ in which the only invalid step is a division by 0. Try to obscure the division by 0.

▶ **Exercise 1.2.** Show that there is a one-to-one correspondence between the set $\mathbf{Q}$ of rational numbers and the following set L of lines. The set L consists of all lines through the origin, except the vertical line $x = 0$, that pass through a nonzero point (a, b) where a and b are integers. (This problem sounds sophisticated, but one word gives the solution!)

The rational number system forms a *field*. A field consists of objects which can be added and multiplied; these operations satisfy the laws we expect. We begin our development by giving the precise definition of a field.

Definition 2.1. A field $\mathbf{F}$ is a mathematical system consisting of a collection of objects and two operations, addition and multiplication, subject to the following axioms.

1) For all x, y in $\mathbf{F}$, we have $x + y = y + x$ and $x * y = y * x$ (the commutative laws for addition and multiplication).

2) For all x, y, t in $\mathbf{F}$, we have $(x + y) + t = x + (y + t)$ and $(x * y) * t = x * (y * t)$ (the associative laws for addition and multiplication).

3) There are distinct distinguished elements 0 and 1 in $\mathbf{F}$ such that, for all x in $\mathbf{F}$, we have $0 + x = x + 0 = x$ and $1 * x = x * 1 = x$ (the existence of additive and multiplicative identities).

4) For each x in $\mathbf{F}$ and each y in $\mathbf{F}$ such that $y \neq 0$, there are $-x$ and $\frac{1}{y}$ in $\mathbf{F}$ such that $x + (-x) = 0$ and $y * \frac{1}{y} = 1$ (the existence of additive and multiplicative inverses).

5) For all x, y, t in $\mathbf{F}$ we have $t * (x + y) = (t * x) + (t * y) = t * x + t * y$ (the distributive law).

For clarity and emphasis we repeat some of the main points. The rational numbers provide a familiar example of a field. In any field we can add, subtract, multiply, and divide as we expect, although we cannot divide by 0. The ability to divide by a nonzero number distinguishes the rational numbers from the integers. In more general settings the ability to divide by a nonzero number distinguishes a field from a commutative ring. Thus every field is a commutative ring but a commutative ring need not be a field.

There are many elementary consequences of the field axioms. It is easy to prove that each element has a unique additive inverse and that each nonzero element has a unique multiplicative inverse, or reciprocal. The proof, left to the reader, mimics our early argument showing that subtraction is possible in $\mathbf{Z}$.

Henceforth we will stop writing $*$ for multiplication; the standard notation of xy for $x * y$ works adequately in most contexts. We also write x^2 instead of xx as usual. Let t be an element in a field. We say that x is a square root of t if $t = x^2$. In a field, taking square roots is not always possible. For example, we shall soon prove that there is no rational square root of 2 and that there is no real square root of -1.

At the risk of boring the reader we prove a few basic facts from the field axioms; the reader who wishes to get more quickly to geometric reasoning could omit the proofs, although writing them out gives one some satisfaction.

Proposition 2.1. *In a field the following laws hold:*

1) $0 + 0 = 0$.

2) *For all x, we have $x0 = 0x = 0$.*

3) $(-1)^2 = (-1)(-1) = 1$.

4) $(-1)x = -x$ *for all x.*

5) *If $xy = 0$ in $\mathbf{F}$, then either $x = 0$ or $y = 0$.*

Proof. Statement 1) follows from setting $x = 0$ in the axiom $0 + x = x$. Statement 2) uses statement 1) and the distributive law to write $0x = (0 + 0)x = 0x + 0x$. By property 4) of Definition 2.1, the object $0x$ has an additive inverse; we add this inverse to both sides of the equation. Using the meaning of additive inverse and then the associative law gives $0 = 0x$. Hence $x0 = 0x = 0$ and 2) holds. Statement 3) is a bit more interesting. We have $0 = 1 + (-1)$ by axiom 4) from Definition 2.1.

Multiplying both sides by -1 and using 2) yields

$$0 = (-1)0 = (-1)(1 + (-1)) = (-1)1 + (-1)^2 = -1 + (-1)^2.$$

Thus $(-1)^2$ is an additive inverse to -1; of course 1 also is. By the uniqueness of additive inverses, we see that $(-1)^2 = 1$. The proof of 4) is similar. Start with $0 = 1 + (-1)$ and multiply by x to get $0 = x + (-1)x$. Thus $(-1)x$ is an additive inverse of x and the result follows by uniqueness of additive inverses. Finally, to prove 5), we assume that $xy = 0$. If $x = 0$, the conclusion holds. If $x \neq 0$, we can multiply by $\frac{1}{x}$ to obtain

$$y = (\frac{1}{x}x)y = \frac{1}{x}(xy) = \frac{1}{x}0 = 0.$$

Thus, if $x \neq 0$, then $y = 0$, and the conclusion also holds. $\square$

We note a point of language, where mathematics usage may differ with common usage. For us, the phrase "either $x = 0$ or $y = 0$" allows the possibility that **both** $x = 0$ and $y = 0$.

Example 2.1. A field with two elements. Let $\mathbf{F}_2$ consist of the two elements 0 and 1. We put $1 + 1 = 0$, but otherwise we add and multiply as usual. Then $\mathbf{F}_2$ is a field.

This example illustrates several interesting things. For example, the object 2 (namely $1 + 1$), can be 0 in a field. This possibility will prevent the quadratic formula from holding in a field for which $2 = 0$. In Theorem 2.1 we will derive the quadratic formula when it is possible to do so.

First we make a simple observation. We have shown that $(-1)^2 = 1$. Hence, when $-1 \neq 1$, it follows that 1 has two square roots, namely ± 1. Can an element of a field have more than two square roots? The answer is no.

Lemma 2.1. *In a field, an element t can have at most two square roots. If x is a square root of t, then so is $-x$, and there are no other possibilities.*

Proof. If $x^2 = t$, then $(-x)^2 = t$ by 3) and 4) of Proposition 2.1. To prove that there are no other possibilities, we assume that both x and y are square roots of t. We then have

$$(2) \qquad\qquad 0 = t - t = x^2 - y^2 = (x - y)(x + y).$$

By 5) of Proposition 2.1, we obtain either $x - y = 0$ or $x + y = 0$. Thus $y = \pm x$ and the result follows. $\square$

The difference of two squares law stating that $x^2 - y^2 = (x - y)(x + y)$ is a gem of elementary mathematics. For example, suppose you are asked to multiply 88 times 92 in your head. You imagine $88 * 92 = (90 - 2)(90 + 2) = 8100 - 4 = 8096$ and impress some audiences. One can also view this algebraic identity for positive integers simply by removing a small square of dots from a large square of dots and rearranging the dots to form a rectangle. The author once used this kind of method when doing volunteer teaching of multiplication to third graders. See Figure 1.1 for a geometric interpretation of the identity in terms of area.

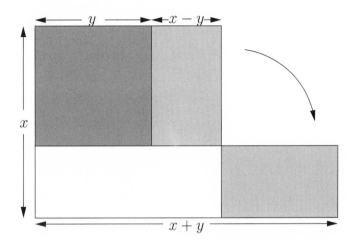

Figure 1.1. Difference of two squares.

We pause to make several remarks about square roots. The first remark concerns a notational convention; the discussion will help motivate the notion of ordered field defined below. The real number system will be defined formally below, and we will prove that positive real numbers have square roots. Suppose $t > 0$. We write $\sqrt{t}$ to denote the *positive* x for which $x^2 = t$. Thus both x and $-x$ are square roots of t, but the notation $\sqrt{t}$ means the positive square root. For the complex numbers, things will be more subtle. We will prove that each nonzero complex number z has two square roots, say $\pm w$, but there is no sensible way to prefer one to the other. We emphasize that the existence of square roots depends on more than the field axioms. Not all positive rational numbers have rational square roots, and hence it must be proved that each positive real number has a square root. The proof requires a limiting process. The quadratic formula, proved next, requires that the expression $b^2 - 4ac$ be a square. In an arbitrary field, the expression $\sqrt{t}$ usually means any x for which $x^2 = t$, but the ambiguity of signs can cause confusion. See Exercise 1.4.

Theorem 2.1. *Let $\mathbf{F}$ be a field. Assume that $2 \neq 0$ in $\mathbf{F}$. For $a \neq 0$ and arbitrary b, c we consider the quadratic equation*

$$(3) \qquad ax^2 + bx + c = 0.$$

Then x solves (3) *if and only if*

$$(4) \qquad x = \frac{-b \pm \sqrt{b^2 - 4ac}}{2a}.$$

If $b^2 - 4ac$ is not a square in $\mathbf{F}$, then (3) *has no solution.*

Proof. The idea of the proof is to *complete the square*. Since both a and 2 are nonzero elements of $\mathbf{F}$, they have multiplicative inverses. We therefore have

$$ax^2 + bx + c = a(x^2 + \frac{b}{a}x) + c = a(x^2 + \frac{b}{a}x + \frac{b^2}{4a^2}) + c - \frac{b^2}{4a}$$

$$(5) \qquad = a(x + \frac{b}{2a})^2 + \frac{4ac - b^2}{4a}.$$

We set (5) equal to 0 and we can easily solve for x. After dividing by a, we obtain

$$(6) \qquad (x + \frac{b}{2a})^2 = \frac{b^2 - 4ac}{4a^2}.$$

The square roots of $4a^2$ are of course $\pm 2a$. Assuming that $b^2 - 4ac$ has a square root in $\mathbf{F}$, we solve (6) for x by first taking the square root of both sides. We obtain

$$(7) \qquad x + \frac{b}{2a} = \pm \frac{\sqrt{b^2 - 4ac}}{2a}.$$

After a subtraction and simplification we obtain (4) from (7). $\qquad\qquad\square$

The reader surely has seen the quadratic formula before. Given a quadratic polynomial with real coefficients, the formula tells us that the polynomial will have no real roots when $b^2 - 4ac < 0$. For many readers the first exposure to complex numbers arises when we introduce square roots of negative numbers in order to use the quadratic formula.

▶ **Exercise 1.3.** Show that additive and multiplicative inverses in a field are unique.

▶ **Exercise 1.4.** A subtlety. Given a field, is the formula

$$\sqrt{u}\sqrt{v} = \sqrt{uv}$$

always valid? In the proof of the quadratic formula, did we use this formula implicitly? If not, what did we use?

Example 2.2. One can completely analyze quadratic equations with coefficients in $\mathbf{F}_2$. The only such equations are $x^2 = 0$, $x^2 + x = 0$, $x^2 + 1 = 0$, and $x^2 + x + 1 = 0$. The first equation has only the solution 0. The second has the two solutions 0 and 1. The third has only the solution 1. The fourth has no solutions. We have given a complete analysis, even though Theorem 2.1 cannot be used in this setting.

Before introducing the notion of ordered field, we give a few other examples of fields. Several of these examples use modular (clock) arithmetic. The phrases *add modulo p* and *multiply modulo p* have the following meaning. Fix a positive integer p, called the *modulus*. Given integers m and n, we add (or multiply) them as usual and then take the remainder upon division by p. The remainder is called the sum (or product) modulo p. This natural notion is familiar to everyone; five hours after nine o'clock is two o'clock; we added modulo twelve. The subsequent examples can be skipped without loss of continuity.

Example 2.3. Fields with finitely many elements. Let p be a prime number, and let $\mathbf{F}_p$ consist of the numbers $0, 1, ..., p-1$. We define addition and multiplication modulo p. Then $\mathbf{F}_p$ is a field.

In Example 2.3, p needs to be a prime number. Property 5) of Proposition 2.1 fails when p is not a prime. We mention without proof that the number of elements in a finite field must be a power of a prime number. Furthermore, for each prime p and positive integer n, there exists a finite field with p^n elements.

▶ **Exercise 1.5.** True or false? Every quadratic equation in $\mathbf{F}_3$ has a solution.

Fields such as $\mathbf{F}_p$ are important in various parts of mathematics and computer science. For us, they will serve only as examples of fields. The most important examples of fields for us will be the real numbers and the complex numbers. To define these fields rigorously will take a bit more effort. We end this section by giving an example of a field built from the real numbers. We will not use this example in the logical development.

Example 2.4. Let K denote the collection of rational functions in one variable x with real coefficients. An element of K can be written $\frac{p(x)}{q(x)}$, where p and q are polynomials, and we assume that q is not the zero polynomial. (We allow $q(x)$ to equal 0 for some x, but not for all x.) We add and multiply such rational functions in the usual way. It is tedious but not difficult to verify the field axioms. Hence K is a field. Furthermore, K contains $\mathbf{R}$ in a natural way; we identify the real number c with the constant rational function $\frac{c}{1}$. As with the rational numbers, many different fractions represent the same element of K. To deal rigorously with such situations, one needs the notion of equivalence relation, discussed in Section 5.

3. Inequalities and ordered fields

Comparing the sizes of a pair of integers or of a pair of rational numbers is both natural and useful. It does not make sense however to compare the sizes of elements in an arbitrary ring or field. We therefore introduce a crucial property shared by the integers $\mathbf{Z}$ and the rational numbers $\mathbf{Q}$. For x, y in either of these sets, it makes sense to say that $x > y$. Furthermore, given the pair x, y, one and only one of three things must be true: $x > y$, $x < y$, or $x = y$. We need to formalize this idea in order to define the real numbers.

Definition 3.1. A field $\mathbf{F}$ is called *ordered* if there is a subset $P \subset \mathbf{F}$, called the *set of positive elements* of $\mathbf{F}$, satisfying the following properties:

    1) For all x, y in P, we have $x + y \in P$ and $xy \in P$ (closure).

    2) For each x in $\mathbf{F}$, one and only one of the following three statements is true: $x = 0$, $x \in P$, $-x \in P$ (trichotomy).

▶ **Exercise 1.6.** Let $\mathbf{F}$ be an ordered field. Show that $1 \in P$.

▶ **Exercise 1.7.** Show that the trichotomy property can be rewritten as follows. For each x, y in $\mathbf{F}$, one and only one of the following three statements is true: $x = y$, $x - y \in P$, $y - x \in P$.

The rational number system is an ordered field; a fraction $\frac{p}{q}$ is positive if and only if p and q have the same sign. Note that q is never 0 and that a rational number is 0 whenever its numerator is 0. It is of course elementary to check in this case that the set P of positive rational numbers is closed under addition and multiplication.

Once the set P of positive elements in a field has been specified, it is easier to work with inequalities than with P. We write $x > y$ if and only if $x - y \in P$. We also use the symbols $x \geq y$, $x \leq y$, $x < y$ as usual. The order axioms then can be written as follows:

1) If $x > 0$ and $y > 0$, then $x + y > 0$ and $xy > 0$.

2) Given $x \in \mathbf{F}$, one and only one of three things holds: $x = 0$, $x > 0$, $x < 0$.

Henceforth we will use inequalities throughout; we mention that these inequalities will compare real numbers. The complex numbers cannot be made into an ordered field. The following lemma about ordered fields does play an important role in our development of the complex number field $\mathbf{C}$.

Lemma 3.1. *Let $\mathbf{F}$ be an ordered field. For each $x \in \mathbf{F}$, we have $x^2 = x * x \geq 0$. If $x \neq 0$, then $x^2 > 0$. In particular, $1 > 0$.*

Proof. If $x = 0$, then $x^2 = 0$ by Proposition 2.1, and the conclusion holdis. If $x > 0$, then $x^2 > 0$ by axiom 1) for an ordered field. If $x < 0$, then $-x > 0$, and hence $(-x)^2 > 0$. By 3) and 4) of Proposition 2.1 we get

$$(8) \qquad\qquad x^2 = (-1)(-1)x^2 = (-x)(-x) = (-x)^2 > 0.$$

Thus, if $x \neq 0$, then $x^2 > 0$. $\qquad\qquad\qquad\qquad\qquad\qquad\qquad\qquad\qquad\qquad\square$

By definition (see Section 3.1), the real number system $\mathbf{R}$ is an ordered field. The following simple corollary motivates the introduction of the complex number field $\mathbf{C}$.

Corollary 3.1. *There is no real number x such that $x^2 = -1$.*

3.1. The completeness axiom for the real numbers. In order to finally define the real number system $\mathbf{R}$, we require the notion of *completeness*. This notion is considerably more advanced than our discussion has been so far. The field axioms allow for algebraic laws, the order axioms allow for inequalities, and the completeness axiom allows for a good theory of *limits*. To introduce this axiom, we recall some basic notions from elementary real analysis. Let $\mathbf{F}$ be an ordered field. Let $S \subset \mathbf{F}$ be a subset. The set S is called *bounded* if there are elements m and M in $\mathbf{F}$ such that

$$m \leq x \leq M$$

for all x in S. The set S is called *bounded above* if there is an element M in $\mathbf{F}$ such that $x \leq M$ for all x in S, and it is called *bounded below* if there is an element m in $\mathbf{F}$ such that $m \leq x$ for all x in S. When these numbers exist, M is called an upper bound for S and m is called a lower bound for S. Thus S is bounded if and only if it is both bounded above and bounded below. The numbers m and M are not generally in S. For example, the set of negative rational numbers is bounded above, but the only upper bounds are 0 or positive numbers.

We can now introduce the completeness axiom for the real numbers. The fundamental notion is that of *least upper bound*. If M is an upper bound for S, then any number larger than M is of course also an upper bound. The term least upper bound means the *smallest* possible upper bound; the concept juxtaposes small and big. The least upper bound α of S is the smallest number that is greater than or equal to any member of S. See Figure 1.2. One cannot prove that such a number exists based on the ordered field axioms; for example, if we work within the realm of rational numbers, the set of x such that $x^2 < 2$ is bounded above, but it has no least upper bound. Mathematicians often use the word *supremum* instead of least upper bound; thus $\sup(S)$ denotes the least upper bound of S. Postulating the existence of least upper bounds as in the next definition uniquely determines the real numbers.

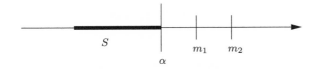

Figure 1.2. Upper bounds.

Definition 3.2. An ordered field $\mathbf{F}$ is *complete* if whenever S is a nonempty subset of $\mathbf{F}$ and S is bounded above, then S has a least upper bound in $\mathbf{F}$.

We could have instead decreed that each nonempty subset of $\mathbf{F}$ that is bounded below has a greatest lower bound (or *infimum*). The two statements are equivalent after replacing S with the set $-S$ of additive inverses of elements of S.

In a certain precise sense, called isomorphism, there is a unique complete ordered field. We will assume uniqueness and get the ball rolling by making the fundamental definition:

Definition 3.3. The real number system $\mathbf{R}$ is the unique complete ordered field.

3.2. What is a natural number? We pause to briefly consider how the natural numbers fit within the real numbers. In our approach, the real number system is taken as the starting point for discussion. From an intuitive point of view we can think of the natural numbers as the set $\{1, 1+1, 1+1+1, ...\}$. To be more precise, we proceed in the following manner.

Definition 3.4. A subset S of $\mathbf{R}$ is called *inductive* if whenever $x \in S$, then $x + 1 \in S$.

Definition 3.5. The set of natural numbers $\mathbf{N}$ is the intersection of all inductive subsets of $\mathbf{R}$ that contain 1.

Thus $\mathbf{N}$ is a subset of $\mathbf{R}$, and $1 \in \mathbf{N}$. Furthermore, if $n \in \mathbf{N}$, then n is an element of every inductive subset of $\mathbf{R}$. Hence $n + 1$ is also an element of every inductive subset of $\mathbf{R}$, and therefore $n+1$ is also in $\mathbf{N}$. Thus $\mathbf{N}$ is itself an inductive set; we could equally well have defined $\mathbf{N}$ to be the *smallest* inductive subset of $\mathbf{R}$ containing 1. As a consequence we obtain the *principle of mathematical induction*:

Proposition 3.1 (Mathematical induction). *Let S be an inductive subset of $\mathbf{N}$ such that $1 \in S$. Then $S = \mathbf{N}$.*

This proposition provides a method of proof, called induction, surely known to many readers. For each $n \in \mathbf{N}$, let P_n be a mathematical statement. To verify that P_n is a true statement for each n, it suffices to show two things: first, P_1 is true; second, for all k, whenever P_k is true, then P_{k+1} is true. The reason is that the set of n for which P_n is true is then an inductive set containing 1; by Proposition 3.1 this set is $\mathbf{N}$.

Figure 1.3. Induction.

▶ **Exercise 1.8.** Apply the principle of mathematical induction to establish the *well-ordering principle*: every nonempty subset of $\mathbf{N}$ contains a least element.

▶ **Exercise 1.9.** It is of course obvious that there is no natural number between 0 and 1. Prove it!

▶ **Exercise 1.10.** For a constant C put $f(x) = x + C$. Find a formula for the composition of f with itself n times. Prove the formula by induction.

▶ **Exercise 1.11.** For nonzero constants A and B, put $f(x) = A(x + B) - B$. Find a formula for the composition of f with itself n times. Prove the formula by induction. Find a short proof by expressing the behavior of f in simple steps.

▶ **Exercise 1.12.** For constants M, C with $M \neq 1$ put $f(x) = Mx + C$. Find a formula for the composition of f with itself n times. Suggestion: Write f in the notation of the previous exercise.

We close this section by proving a precise statement to the effect that many small things make a big thing. This seemingly evident but yet surprisingly subtle property of $\mathbf{R}$, as stated in Proposition 3.2, requires the completeness axiom for its proof. The proposition does not hold in all ordered fields. In other words, there exist ordered fields $\mathbf{F}$ with the following striking property: $\mathbf{F}$ contains the natural numbers, but it also contains *super numbers*, namely elements larger than any natural number. For the real numbers, however, things are as we believe. The natural numbers are an unbounded subset of the real numbers.

Proposition 3.2 (Archimedean property). *Given positive real numbers x and ϵ, there is a positive integer n such that $n\epsilon > x$. Equivalently, given $y > 0$, there is an $n \in \mathbf{N}$ such that $\frac{1}{n} < y$.*

Proof. If the first conclusion were false, then every natural number would be bounded above by $\frac{x}{\epsilon}$. If the second conclusion were false, then every natural number would be bounded above by $\frac{1}{y}$. Thus, in either case, $\mathbf{N}$ would be bounded above. We prove otherwise. If $\mathbf{N}$ were bounded above, then by the completeness axiom $\mathbf{N}$ would have a least upper bound K. But then $K - 1$ would not be an upper bound, and hence we could find an integer n with $K - 1 < n \leq K$. But then $K < n + 1$; since $n + 1 \in \mathbf{N}$, we contradict K being an upper bound. Thus $\mathbf{N}$ is unbounded above and the Archimedean property follows. $\qquad\square$

▶ **Exercise 1.13.** Type "Non-Archimedean Ordered Field" into an internet search engine and see what you find. Then try to understand one of the examples.

3.3. Limits. Completeness in the sense of Definition 3.2 (for Archimedean ordered fields) is equivalent to a notion involving limits of Cauchy sequences. See Remark 3.1. We will carefully discuss these definitions from a calculus or beginning real analysis course. First we remind the reader of some elementary properties of the absolute value function. We gain intuition by thinking in terms of distance.

Definition 3.6. For $x \in \mathbf{R}$, we define $|x|$ by $|x| = x$ if $x \geq 0$ and $|x| = -x$ if $x < 0$. Thus $|x|$ represents the distance between x and 0. In general, we define the *distance* $\delta(x, y)$ between real numbers x and y by

$$\delta(x, y) = |x - y|.$$

▶ **Exercise 1.14.** Show that the absolute value function on $\mathbf{R}$ satisfies the following properties:

1) $|x| \geq 0$ for all $x \in \mathbf{R}$, and $|x| = 0$ if and only if $x = 0$.

2) $-|x| \leq x \leq |x|$ for all $x \in \mathbf{R}$.

3) $|x + y| \leq |x| + |y|$ for all $x, y \in \mathbf{R}$ (the triangle inequality).

4) $|a - c| \leq |a - b| + |b - c|$ for all $a, b, c \in \mathbf{R}$ (second form of the triangle inequality).

▶ **Exercise 1.15.** Why are properties 3) and 4) of the previous exercise called triangle inequalities?

We make several comments about Exercise 1.14. First of all, one can prove property 3) in two rather different ways. One way starts with property 2) for x and y and adds the results. Another way involves squaring. Property 4) is crucial because of its interpretation in terms of distances. Mathematicians have abstracted these properties of the absolute value function and introduced the concept of a *metric space*. See Section 6.

We recall that a sequence $\{x_n\}$ of real numbers is a function from $\mathbf{N}$ to $\mathbf{R}$. The real number x_n is called the n-th term of the sequence. The notation $x_1, x_2, ..., x_n, ...$, where we list the terms of the sequence in order, amounts to listing the values of the function. Thus $x : \mathbf{N} \to \mathbf{R}$ is a function, and we write x_n instead of $x(n)$. The intuition gained from this alteration of notation is especially valuable when discussing limits.

Definition 3.7. Let $\{x_n\}$ be a sequence of real numbers. Assume $L \in \mathbf{R}$.

- LIMIT. We say that "the limit of x_n is L" or that "x_n converges to L", and we write $\lim_{n \to \infty} x_n = L$ if the following statement holds: For all $\epsilon > 0$, there is an $N \in \mathbf{N}$ such that $n \geq N$ implies $|x_n - L| < \epsilon$.

- CAUCHY. We say that $\{x_n\}$ is a *Cauchy sequence* if the following statement holds: For all $\epsilon > 0$, there is an $N \in \mathbf{N}$ such that $m, n \geq N$ implies $|x_m - x_n| < \epsilon$.

When there is no real number L for which $\{x_n\}$ converges to L, we say that $\{x_n\}$ *diverges*.

The definition of the limit demands that the terms eventually get arbitrarily close to a given L. The definition of a Cauchy sequence states that the terms of the sequence eventually get arbitrarily close to each other. The most fundamental result in real analysis is that a sequence of real numbers converges if and only if it is a Cauchy sequence. The word *complete* has several similar uses in mathematics; it often refers to a metric space in which being Cauchy is a necessary and sufficient condition for convergence of a sequence. See Section 6. The following subtle remark indicates a slightly different way one can define the real numbers.

Remark 3.1. Consider an ordered field $\mathbf{F}$ satisfying the Archimedean property. In other words, given positive elements x and y, there is an integer n such that y added to itself n times exceeds x. Of course we write ny for this sum. It is possible to consider limits and Cauchy sequences in $\mathbf{F}$. Suppose that each Cauchy sequence in $\mathbf{F}$ has a limit in $\mathbf{F}$. One can then derive the least upper bound property, and $\mathbf{F}$ must be the real numbers $\mathbf{R}$. Hence we could give the definition of the real number system by decreeing that $\mathbf{R}$ is an ordered field satisfying the Archimedean property and that $\mathbf{R}$ is complete in the sense of Cauchy sequences.

We return to the real numbers. A sequence $\{x_n\}$ of real numbers is *bounded* if and only if its set of values is a bounded subset of $\mathbf{R}$. A convergent sequence must of course be bounded; with finitely many exceptions all the terms are within distance 1 from the limit. Similarly a Cauchy sequence must be bounded; with finitely many exceptions all the terms are within distance 1 of some particular x_N.

Proving that a convergent sequence must be Cauchy uses what is called an $\frac{\epsilon}{2}$ argument. Here is the idea: if the terms are eventually within distance $\frac{\epsilon}{2}$ of some limit L, then they are eventually within distance ϵ of each other. Proving the converse assertion is much more subtle; somehow one must find a candidate for the limit just knowing that the terms are close to each other. See for example [**8, 20**]. The proofs rely on the notion of subsequence, which we define now, but which we do not use meaningfully until Chapter 8. Let $\{x_n\}$ be a sequence of real numbers and let $k \to n_k$ be an increasing function. We write $\{x_{n_k}\}$ for the subsequence of $\{x_n\}$ whose k-th term is x_{n_k}. The proof that a Cauchy sequence converges amounts to first finding a convergent subsequence and then showing that the sequence itself converges to the same limit.

We next prove a basic fact that often allows us to determine convergence of a sequence without knowing the limit in advance. A sequence $\{x_n\}$ is called *nondecreasing* if, for each n, we have $x_{n+1} \geq x_n$. It is called *nonincreasing* if, for each n, we have $x_{n+1} \leq x_n$. It is called *monotone* if it is either nonincreasing or nondecreasing. The following fundamental result, illustrated by Figure 1.4, will get used occasionally in this book. It can be used also to establish that a Cauchy sequence of real numbers has a limit.

Proposition 3.3. *A bounded monotone sequence of real numbers has a limit.*

Proof. We claim that a nondecreasing sequence converges to its least upper bound (supremum) and that a nonincreasing sequence converges to its greatest lower bound (infimum). We prove the first, leaving the second to the reader. Suppose for all n we have

$$x_1 \leq \dots \leq x_n \leq x_{n+1} \leq \dots \leq M.$$

Let α be the least upper bound of the set $\{x_n\}$. Then, given $\epsilon > 0$, the number $\alpha - \epsilon$ is not an upper bound, and hence there is some x_N with $\alpha - \epsilon < x_N \leq \alpha$. By the nondecreasing property, if $n \geq N$, then

(9) $$\alpha - \epsilon < x_N \leq x_n \leq \alpha < \alpha + \epsilon.$$

But (9) yields $|x_n - \alpha| < \epsilon$ and hence provides us with the needed N in the definition of the limit. Thus $\lim_{n \to \infty}(x_n) = \alpha$. $\qquad\square$

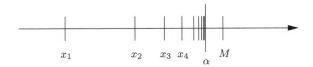

Figure 1.4. Monotone convergence.

Remark 3.2. Let $\{x_n\}$ be a monotone sequence of real numbers. Then $\{x_n\}$ converges if and only if it is bounded. Proposition 3.3 guarantees that it converges if it is bounded. Since a convergent sequence must be bounded, the converse holds as well. Monotonicity is required; for example, the sequence $(-1)^n$ is bounded but it does not converge.

The next few pages provide the basic real analysis needed as background material. In particular the material on square roots is vital to the development.

▶ **Exercise 1.16.** Finish the proof of Proposition 3.3; in other words, show that a nonincreasing bounded sequence converges to its greatest lower bound.

▶ **Exercise 1.17.** If c is a constant and $\{x_n\}$ converges, prove that $\{cx_n\}$ converges. Try to arrange your proof such that the special case $c = 0$ need not be considered separately. Prove that the sum and product of convergent sequences are convergent.

▶ **Exercise 1.18.** Assume $\{x_n\}$ converges to 0 and that $\{y_n\}$ is bounded. Prove that their product converges to 0.

An extension of the notion of limit of sequence is often useful in real analysis. We pause to introduce the idea and refer to [**20**] for applications and considerably more discussion. When S is a bounded and nonempty subset of $\mathbf{R}$, we write as usual $\inf(S)$ for the greatest lower bound of S and $\sup(S)$ for the least upper bound of S. Let now $\{x_n\}$ be a bounded sequence of real numbers. For each k, consider the set $X_k = \{x_n : n \geq k\}$. Then these sets are bounded as well. Furthermore the bounded sequence of real numbers defined by $\inf(X_k)$ is nondecreasing and the bounded sequence of real numbers $\sup(X_k)$ is nonincreasing. By the monotone convergence theorem these sequences necessarily have limits, called $\lim\inf(x_n)$ and $\lim\sup(x_n)$. These limits are equal **if and only if** $\lim(x_n)$ exists, in which case all three values are the same. By contrast, let $x_n = (-1)^n$. Then $\lim \inf(x_n) = -1$ and $\lim \sup(x_n) = 1$. Occasionally in the subsequent discussion we can replace limit by lim sup and things still work.

Next we turn to the concept of continuity, which we also define in terms of sequences.

Definition 3.8. Let $f : \mathbf{R} \to \mathbf{R}$ be a function. Then f is *continuous at a* if whenever $\{x_n\}$ is a sequence and $\lim_{n\to\infty} x_n = a$, then $\lim_{n\to\infty} f(x_n) = f(a)$. Also, f is *continuous on a set S* if it is continuous at each point of the set. When S is $\mathbf{R}$ or when S is understood from the context to be the domain of f, we usually say "f is continuous" rather than the longer phrase "f is continuous on S".

▶ **Exercise 1.19.** Prove that the sum and product of continuous functions are continuous. If c is a constant and f is continuous, prove that cf is continuous.

▶ **Exercise 1.20.** Prove that f is continuous at a if and only if the following holds. For each $\epsilon > 0$, there is a $\delta > 0$ such that $|x - a| < \delta$ implies $|f(x) - f(a)| < \epsilon$.

We close this section by showing how the completeness axiom impacts the existence of square roots. First we recall the standard fact that there is no rational square root of 2, by giving a somewhat unusual proof. See Exercise 1.22 for a compelling generalization. These proofs are based on inequalities. For example, the order axioms yield the following: $0 < a < b$ implies $0 < a^2 < ab < b^2$; we use such inequalities without comment below.

Proposition 3.4. *There is no rational number whose square is 2.*

Proof. Seeking a contradiction, we suppose that there are integers m, n such that $(\frac{m}{n})^2 = 2$. We may assume that m and n are positive. Of all such representations we may assume that we have chosen the one for which n is the smallest possible positive integer. The equality $m^2 = 2n^2$ implies the inequality $2n > m > n$. Now we compute

$$(10) \qquad \frac{m}{n} = \frac{m(m-n)}{n(m-n)} = \frac{m^2 - mn}{n(m-n)} = \frac{2n^2 - mn}{n(m-n)} = \frac{2n - m}{m - n}.$$

Thus $\frac{2n-m}{m-n}$ is also a square root of 2. Since $0 < m - n < n$, formula (10) provides a second way to write the fraction $\frac{m}{n}$; the second way has a positive denominator, smaller than n. We have therefore contradicted our choice of n. Hence there is no rational number whose square is 2. □

Although there is no *rational* square root of 2, we certainly believe that a positive *real* square root of 2 exists. For example, the length of the diagonal of the unit square should be $\sqrt{2}$. We next prove, necessarily relying on the completeness axiom, that each positive real number has a square root.

Theorem 3.1. *If $t \in \mathbf{R}$ and $t \geq 0$, then there is an $x \in \mathbf{R}$ with $x^2 = t$.*

Proof. This proof is somewhat sophisticated and can be omitted on first reading. If $t = 0$, then t has the square root 0. Hence we may assume that $t > 0$. Let S denote the set of real numbers x such that $x^2 < t$. This set is nonempty, because $0 \in S$. We claim that $M = \max(1, t)$ is an upper bound for S. To check the claim, we note first that $x^2 < 1$ implies $x < 1$, because $x \geq 1$ implies $x^2 \geq 1$. Therefore if $t < 1$, then 1 is an upper bound for S. On the other hand, if $t \geq 1$, then $t \leq t^2$. Therefore $x^2 < t$ implies $x^2 \leq t^2$ and hence $x \leq t$. Therefore in this case t is an upper bound for S. In either case, S is bounded above by M and is nonempty. By the completeness axiom, S has a least upper bound α. We claim that $\alpha^2 = t$.

To prove the claim, we use the trichotomy property. We will rule out the cases $\alpha^2 < t$ and $\alpha^2 > t$. In each case we use the Archimedean property to find a positive integer n whose reciprocal is sufficiently small. Then we can add or subtract $\frac{1}{n}$ to α and obtain a contradiction. Here are the details. If $\alpha^2 > t$, then Proposition 3.2 guarantees that we can find an integer n such that

$$\frac{2\alpha}{n} < \alpha^2 - t.$$

We then have

$$(\alpha - \frac{1}{n})^2 = \alpha^2 - \frac{2\alpha}{n} + \frac{1}{n^2} > \alpha^2 - \frac{2\alpha}{n} > t.$$

Thus $\alpha - \frac{1}{n}$ is an upper bound for S, but it is smaller than α. We obtain a contradiction. Suppose next that $\alpha^2 < t$. We can find $n \in \mathbf{N}$ (Exercise 1.21) such that

(11)
$$\frac{2\alpha n + 1}{n^2} < t - \alpha^2.$$

This time we obtain

$$(\alpha + \frac{1}{n})^2 = \alpha^2 + \frac{2\alpha}{n} + \frac{1}{n^2} < t.$$

Since $\alpha + \frac{1}{n}$ is bigger than α and yet it is also an upper bound for S, again we obtain a contradiction. By trichotomy we must therefore have $\alpha^2 = t$. $\qquad \square$

The kind of argument used in the proof of Theorem 3.1 epitomizes proofs in basic real analysis. In this setting one cannot prove an equality by algebraic reasoning; one requires the completeness axiom and analytic reasoning.

▶ **Exercise 1.21.** For $t - \alpha^2 > 0$, prove that there is an $n \in \mathbf{N}$ such that (11) holds.

▶ **Exercise 1.22.** Mimic the proof of Proposition 3.4 to prove the following statement. If k is a positive integer, then the square root of k must be either an integer or an irrational number. Suggestion: Multiply $\frac{m}{n}$ by $\frac{m-nq}{m-nq}$ for a suitable integer q.

4. The complex numbers

We are finally ready to introduce the complex numbers $\mathbf{C}$. The equation $x^2 + 1 = 0$ will have two solutions in $\mathbf{C}$. Once we allow a solution to this equation, we find via the quadratic formula and Lemma 4.1 below that we can solve all quadratic polynomial equations. With deeper work, we can solve any (nonconstant) polynomial equation over $\mathbf{C}$. We will prove this result, called the *fundamental theorem of algebra*, in Chapter 8.

Our first definition of $\mathbf{C}$ arises from algebraic reasoning. As usual, we write $\mathbf{R}^2$ for the set of ordered pairs (x, y) of real numbers. To think geometrically, we identify the point (x, y) with the arrow from the origin $(0, 0)$ to the point (x, y). We know how to add vectors; hence we define

(12)
$$(x, y) + (a, b) = (x + a, y + b).$$

This formula amounts to adding vectors in the usual geometric manner. See Figure 1.5. More subtle is our definition of multiplication

(13)
$$(x, y) * (a, b) = (xa - yb, xb + ya).$$

Let us temporarily write $\mathbf{0}$ for $(0, 0)$ and $\mathbf{1}$ for $(1, 0)$. We claim that the operations in equations (12) and (13) turn $\mathbf{R}^2$ into a field.

We must first verify that both addition and multiplication are commutative and associative. The verifications are rather trivial, especially for addition:

$$(x, y) + (a, b) = (x + a, y + b) = (a + x, b + y) = (a, b) + (x, y),$$

$$((x, y) + (a, b)) + (s, t) = (x + a, y + b) + (s, t) = (x + a + s, y + b + t)$$
$$= (x, y) + (a + s, b + t) = (x, y) + ((a, b) + (s, t)).$$

Here are the computations for multiplication:

$$(x, y) * (a, b) = (xa - yb, xb + ya) = (ax - by, ay + bx) = (a, b) * (x, y),$$

$$((x, y) * (a, b)) * (s, t) = (xa - yb, xb + ya) * (s, t)$$
$$= (xas - bys - txb - tya, xbt + yat - (xas - bys)) = (x, y) * ((a, b) * (s, t)).$$

We next verify that $\mathbf{0}$ and $\mathbf{1}$ have the desired properties.

$$(x, y) + (0, 0) = (x, y),$$

$$(x, y) * (1, 0) = (x1 - y0, x0 + y1) = (x, y).$$

The additive inverse of (x, y) is easily checked to be $(-x, -y)$. When $(x, y) \neq (0, 0)$, the multiplicative inverse of (x, y) is easily checked to be

(14)
$$\frac{\mathbf{1}}{(x, y)} = \left(\frac{x}{x^2 + y^2}, \frac{-y}{x^2 + y^2}\right).$$

Checking the distributive law is not hard, but it is tedious and left to the reader in Exercise 1.23.

These calculations provide the starting point for discussion.

Theorem 4.1. *Formulas* (12) *and* (13) *make* $\mathbf{R}^2$ *into a field.*

The verification of the field axioms given above is rather dull and uninspired. We do note, however, that $(-1, 0)$ is the additive inverse of $(1, 0) = \mathbf{1}$ and that $(0, 1) * (0, 1) = (-1, 0)$. Hence there is a square root of -1 in this field.

The ordered pair notation for elements is a bit awkward. We wish to give two alternative definitions of $\mathbf{C}$ where things are more elegant.

What have we done so far? Our first definition of $\mathbf{C}$ as pairs of real numbers gave an unmotivated recipe for multiplication; it seems almost a fluke that we obtain a field using this definition. Furthermore computations seem clumsy. A more appealing approach begins by introducing a formal symbol $\mathbf{i}$ and defining $\mathbf{C}$

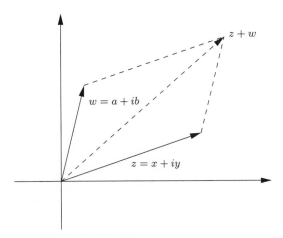

Figure 1.5. Addition of complex numbers.

to be the set of expressions of the form $a + \mathbf{i}b$ for real numbers a, b. We add and multiply as expected, using the distributive law; then we set $\mathbf{i}^2$ equal to -1. Thus

$$(15) \qquad (x + \mathbf{i}y) + (a + \mathbf{i}b) = (x + a) + \mathbf{i}(y + b),$$

$$(16) \qquad (x + \mathbf{i}y) * (a + \mathbf{i}b) = xa + \mathbf{i}(ya + xb) + \mathbf{i}^2(yb) = (xa - yb) + \mathbf{i}(ya + xb).$$

Equations (15) and (16) give the same results as (12) and (13). While this new approach is more elegant, it makes some readers feel uneasy. After all, we are assuming the existence of an object, namely $0 + \mathbf{i}1$, whose square is -1. In the first approach we never assume the existence of such a thing, but such a thing does exist: the square of $(0, 1)$ is $(-1, 0)$, which is the additive inverse of $(1, 0)$.

The reader will be on safe logical ground if he or she regards the above paragraph as an abbreviation for the previous discussion. In the next section we will give two additional equivalent ways of defining **C**.

▶ **Exercise 1.23.** Prove the distributive law for addition and multiplication, as defined in (12) and (13). Do the same using (15) and (16). Compare.

The next lemma reveals a crucial difference between **R** and **C**.

Lemma 4.1. *The complex numbers do not form an ordered field.*

Proof. Assume that a positive subset P exists. By Lemma 3.1, each nonzero square is in P. Since $1^2 = 1$ and $i^2 = -1$, both 1 and -1 are squares and hence must be positive, contradicting 2) of Definition 3.1.   □

5. Alternative definitions of C

In this section we discuss alternative approaches to defining **C**. We use some basic ideas from linear and abstract algebra that might be new to many students. The primary purpose of this section is to assuage readers who find the rules (12) and (13) unappealing but who find the rules (15) and (16) dubious, because we introduced an object $\mathbf{i}$ whose square is -1. The first approach uses matrices of real numbers, and it conveys significant geometric information. The second approach fully justifies starting with (15) and (16) and it provides a quintessential example of what mathematicians call a *quotient space*.

A matrix approach to C. The matrix definition of $\mathbf{C}$ uses two-by-two matrices of real numbers and some of the ideas are crucial to subsequent developments. In this approach we think of $\mathbf{C}$ as the set of two-by-two matrices of the form (18), thereby presaging the Cauchy-Riemann equations which will appear throughout the book. In some sense we identify a complex number with the operation of multiplication by that complex number. This approach is especially useful in complex geometry.

We can regard a complex number as a special kind of linear transformation of $\mathbf{R}^2$. A general linear transformation $(x, y) \to (ax + cy, bx + dy)$ is given by a two-by-two matrix M of real numbers:

$$(17) \qquad\qquad M = \begin{pmatrix} a & c \\ b & d \end{pmatrix}.$$

A complex number will be a special kind of two-by-two matrix. Given a pair of real numbers a, b and motivated by (13), we consider the mapping $L : \mathbf{R}^2 \to \mathbf{R}^2$ defined by

$$L(x, y) = (ax - by, bx + ay).$$

The matrix representation (in the standard basis) of this linear mapping L is the two-by-two matrix

$$(18) \qquad\qquad \begin{pmatrix} a & -b \\ b & a \end{pmatrix}.$$

We say that a two-by-two matrix of real numbers satisfies the *Cauchy-Riemann equations* if it has the form (18). A real linear transformation from $\mathbf{R}^2$ to itself whose matrix representation satisfies (18) corresponds to a *complex linear* transformation from $\mathbf{C}$ to itself, namely multiplication by $a + ib$.

In this approach we *define* a complex number to be a two-by-two matrix (of real numbers) satisfying the Cauchy-Riemann equations. We add and multiply matrices in the usual manner. We then have an additive identity $\mathbf{0}$, a multiplicative identity $\mathbf{1}$, an analogue of $\mathbf{i}$, and inverses of nonzero elements, defined as follows:

$$(19) \qquad\qquad \mathbf{0} = \begin{pmatrix} 0 & 0 \\ 0 & 0 \end{pmatrix},$$

$$(20) \qquad\qquad \mathbf{1} = \begin{pmatrix} 1 & 0 \\ 0 & 1 \end{pmatrix},$$

$$(21) \qquad\qquad \mathbf{i} = \begin{pmatrix} 0 & -1 \\ 1 & 0 \end{pmatrix}.$$

If a and b are not both 0, then $a^2 + b^2 > 0$. Hence in this case the matrix

$$(22) \qquad\qquad \begin{pmatrix} \frac{a}{a^2+b^2} & \frac{b}{a^2+b^2} \\ \frac{-b}{a^2+b^2} & \frac{a}{a^2+b^2} \end{pmatrix}$$

makes sense and satisfies the Cauchy-Riemann equations. Note that

$$(23) \qquad \begin{pmatrix} a & -b \\ b & a \end{pmatrix} \begin{pmatrix} \frac{a}{a^2+b^2} & \frac{b}{a^2+b^2} \\ \frac{-b}{a^2+b^2} & \frac{a}{a^2+b^2} \end{pmatrix} = \begin{pmatrix} 1 & 0 \\ 0 & 1 \end{pmatrix} = \mathbf{1}.$$

Thus (22) yields the formula $\frac{a-ib}{a^2+b^2}$ for the reciprocal of the nonzero complex number $a + ib$, expressed instead in matrix notation.

Thus **C** can be defined to be the set of two-by-two matrices satisfying the Cauchy-Riemann equations. Addition and multiplication are defined as usual for matrices. The additive identity **0** is given by (19) and the multiplicative identity **1** is given by (20). The resulting mathematical system is a field, and the element **i** defined by (21) satisfies $\mathbf{i}^2 + \mathbf{1} = \mathbf{0}$. This method of defining **C** should appease readers who on philosophical grounds question the existence of complex numbers.

▶ **Exercise 1.24.** Show that the square of the matrix in (21) is the negative of the matrix in (20); in other words, show that $\mathbf{i}^2 = -\mathbf{1}$.

▶ **Exercise 1.25.** Suppose $a^2 + b^2 = 1$ in (18). What is the geometric meaning of multiplication by L?

▶ **Exercise 1.26.** Suppose $b = 0$ in (18). What is the geometric meaning of multiplication by L?

▶ **Exercise 1.27.** Show that there are no real numbers x and y such that

$$\frac{1}{x} + \frac{1}{y} = \frac{1}{x+y}.$$

Show on the other hand that there are complex numbers z and w such that

$$(24) \qquad \frac{1}{z} + \frac{1}{w} = \frac{1}{z+w}.$$

Describe all pairs (z, w) satisfying (24).

▶ **Exercise 1.28.** Describe all pairs A and B of two-by-two matrices of real numbers for which A^{-1} and B^{-1} exist and

$$A^{-1} + B^{-1} = (A + B)^{-1}.$$

Remark 5.1. Such pairs of n-by-n matrices exist if and only if n is even; the reason is intimately connected with complex analysis.

An algebraic definition of C. We next describe **C** as a *quotient space*. This approach allows us to regard a complex number as an expression $a + ib$, where $i^2 = -1$, as we wish to do. We will therefore define **C** in terms of the polynomial ring divided by an ideal. The reader may skip this section without loss of understanding.

First we recall the general notion of an *equivalence relation*. Let S be a set. We can think of an equivalence relation on S as being defined via a symbol $\cong$. We decree that certain pairs of elements $s, t \in S$ are *equivalent*; if so, we write $s \cong t$. The following three axioms must hold:

- For all $s \in S$, $s \cong s$ (reflexivity).
- For all $s, t \in S$, $s \cong t$ if and only if $t \cong s$ (symmetry).
- For all $s, t, u \in S$, $s \cong t$ and $t \cong u$ together imply $s \cong u$ (transitivity).

Given an equivalence relation $\cong$ on S, we partition S into *equivalence classes*. All the elements in a single equivalence class are equivalent, and no other member of S is equivalent to any of these elements. We have already seen two elementary examples. First, fractions $\frac{a}{b}$ and $\frac{c}{d}$ are equivalent if and only if they represent the same real number, that is, if and only if $ad = bc$. Thus a rational number may be regarded as an equivalence class of pairs of integers. Second, when doing arithmetic modulo p, we regard two integers as being in the same equivalence class if their difference is divisible by p.

▶ **Exercise 1.29.** The precise definition of modular arithmetic involves equivalence classes; we add and multiply equivalence classes (rather than numbers). Show that addition and multiplication modulo p are well-defined concepts. In other words, do the following. Assume m_1 and m_2 are in the same equivalence class modulo p and that n_1 and n_2 are also in the same equivalence class (not necessarily the same class m_1 and m_2 are in). Show that $m_1 + n_1$ and $m_2 + n_2$ are in the same equivalence class modulo p. Do the same for multiplication.

▶ **Exercise 1.30.** Let S be the set of students at a college. For $s, t \in S$, consider the relation $s \cong t$ if s and t take a class together. Is this relation an equivalence relation?

Let $\mathbf{R}[t]$ denote the collection of polynomials in one variable, with real coefficients. An element p of $\mathbf{R}[t]$ can be written

$$p = \sum_{j=0}^{d} a_j t^j,$$

where $a_j \in \mathbf{R}$. Notice that the sum is finite. Unless all the a_j are 0, there is a largest d for which $a_j \neq 0$. This number d is called the *degree* of the polynomial. When all the a_j equal 0, we call the resulting polynomial the *zero polynomial* and agree that it has no degree. (In some contexts, one assigns the symbol $-\infty$ to be the degree of the zero polynomial.) The sum and the product of polynomials are defined as in high school mathematics. In many ways $\mathbf{R}[t]$ resembles the integers $\mathbf{Z}$. Each is a commutative ring under the operations of sum and product. Unique factorization into irreducible elements holds in both settings, and the division algorithm works the same as well. See [**4**] or [**8**] for more details. Given polynomials p and g, we say that p is a *multiple* of g, or equivalently that g *divides* p, if there is a polynomial q with $p = gq$.

The polynomial $1 + t^2$ is irreducible, in the sense that it cannot be written as a product of two polynomials, each of lower degree, with real coefficients. The set I of polynomials divisible by $1 + t^2$ is called the *ideal generated by* $1 + t^2$. Given two polynomials p, q, we say that they are equivalent modulo I if $p - q \in I$, in other words, if $p - q$ is divisible by $1 + t^2$. We observe that the three properties of an equivalence relation hold:

- For all p, $p \cong p$.
- For all p, q, $p \cong q$ if and only if $q \cong p$.
- If $p \cong q$ and $q \cong r$, then $p \cong r$.

This equivalence relation partitions the set $\mathbf{R}[t]$ into equivalence classes; the situation is strikingly similar to modular arithmetic. Given a polynomial $p(t)$, we use the division algorithm to write $p(t) = q(t)(1 + t^2) + r(t)$, where the remainder r has degree at most one. Thus $r(t) = a + bt$ for some a, b, and this r is the unique first-degree polynomial equivalent to p. In the case of modular arithmetic we used the remainder upon division by the modulus; here we use the remainder upon division by $t^2 + 1$.

▶ **Exercise 1.31.** Verify the transitivity property of equivalence modulo I.

Standard notation in algebra writes $\mathbf{R}[t]/(1 + t^2)$ for the set of equivalence classes. We can add and multiply in $\mathbf{R}[t]/(1 + t^2)$. As usual, the sum (or product) of equivalence classes P and Q is defined to be the equivalence class of the sum $p + q$ (or the product pq) of members; the result is independent of the choice. An equivalence class then can be identified with a polynomial $a + bt$, and the sum and product of equivalence classes satisfies (15) and (16). In this setting we define $\mathbf{C}$ as the collection of equivalence classes with this natural sum and product:

$$(25) \qquad\qquad \mathbf{C} = \mathbf{R}[t]/(1 + t^2).$$

Definition (25) allows us to set $t^2 = -1$ whenever we encounter a term of degree at least two. The irreducibility of $t^2 + 1$ matters. If we form $\mathbf{R}[t]/(p(t))$ for a reducible polynomial p, then the resulting object will not be a field. The reason is precisely parallel to the situation with modular arithmetic. If we consider $\mathbf{Z}/(n)$, then we get a field (written $\mathbf{F}_n$) if and only if n is prime.

▶ **Exercise 1.32.** Show that $\mathbf{R}[t]/(t^3 + 1)$ is not a field.

▶ **Exercise 1.33.** A polynomial $\sum_{k=0}^{d} c_k t^k$ in $\mathbf{R}[t]$ is equivalent to precisely one polynomial of the form $A + Bt$ in the quotient space. What is $A + Bt$ in terms of the coefficients c_k?

▶ **Exercise 1.34.** Prove the division algorithm in $\mathbf{R}[t]$. In other words, given polynomials p and g, with g not the zero polynomial, show that one can write $p = qg + r$ where either $r = 0$ or the degree of r is less than the degree of g. Show that q and r are uniquely determined by p and g.

▶ **Exercise 1.35.** For any polynomial p and any x_0, show that there is a polynomial q such that $p(x) = (x - x_0)q(x) + p(x_0)$.

6. A glimpse at metric spaces

Both the real number system and the complex number system provide intuition for the general notion of a metric space. This section can be omitted without impacting the logical development, but it should appeal to some readers.

Definition 6.1. Let X be a set. A *distance function* on X is a function $\delta : X \times X \to \mathbf{R}$ such that the following hold:

    1) $\delta(x, y) \geq 0$ for all $x, y \in X$ (distances are nonnegative).

    2) $\delta(x, y) = 0$ if and only if $x = y$ (distinct points have positive distance between them; a point has 0 distance to itself).

3) $\delta(x, y) = \delta(y, x)$ for all $x, y \in X$ (the distance from x to y is the same as the distance from y to x).

4) $\delta(x, w) \leq \delta(x, y) + \delta(y, w)$ for all $x, y, w \in X$ (the triangle inequality).

Definition 6.2. Let X be a set, and let δ be a distance function on X. The pair (X, δ) is called a *metric space*.

Mathematicians sometimes write that "the distance function δ makes the set X into a metric space". That statement is not completely precise. It ignores a somewhat pedantic point: the set X is not the metric space; a metric space consists of the set and the distance function. Many different distance functions are possible for most sets X.

Sometimes we use the word *metric* on its own to mean the distance function on a metric space. The most intuitive example is the real number system; putting $\delta(x, y) = |x - y|$ makes **R** into a metric space. There are many other possibilities for metrics on **R**; for example, putting $\delta(x, y)$ equal to 1 whenever $x \neq y$ gives another possible distance function. In Chapter 2 we formally define the absolute value function for complex numbers. Then **C** becomes a metric space when we put $\delta(z, w) = |z - w|$. Again, many other distance functions exist. We mention these ideas now for primarily one reason. The basic concepts involving sequences and limits can be developed in the metric space setting. The concept of completeness is then based upon Cauchy sequences; a metric space (X, δ) is *complete* if and only if every Cauchy sequence in X has a limit in X. Given a metric space that is not complete in this sense, it is possible to enlarge it by including all limits of Cauchy sequences. This approach provides one method for defining the real numbers in terms of the rational numbers. Recall that our development assumes the existence of the real numbers and defines the rational numbers as a particular subset of the real numbers.

In a metric space there are notions of *open* and *closed* balls. Given $p \in X$, we write $B_r(p)$ for the set of points whose distance to p is less than r; we call $B_r(p)$ the open ball of radius r about p. The closed ball also includes points whose distance to p equals r. Depending on the metric δ, these sets might not resemble our usual geometric picture of balls. A subset S of a metric space is *open* if, for each $p \in S$, there is an $\epsilon > 0$ such that $B_\epsilon(p) \subset S$. In most situations what counts is the collection of open subsets of a metric space. It is possible and also appealing to define all the basic concepts (limit, continuous function, bounded, compact, connected, etc.) in terms of the collection of open sets. The resulting subject is called *point set topology*. In order to keep this book at an elementary level, we will not do so; instead we rely on the metric in **C** defined by $\delta(z, w) = |z - w|$. The distance between points in this metric equals the usual Euclidean distance between them.

Occasionally we will require that an open set be connected, and hence we give the definition here. In Definition 6.3 and a few other times in this book, the symbol $\varnothing$ denotes the empty set. The empty set is open; there are no points in $\varnothing$, and hence the definition of open is satisfied, albeit vacuously.

Definition 6.3. Let (X, δ) be a metric space. A subset S of X is called connected if the following holds: whenever $S = A \cup B$, where A and B are open and $A \cap B = \varnothing$, then either $A = \varnothing$ or $B = \varnothing$.

In Chapter 6 we will use the following fairly simple result. In it and in the subsequent comment we assume $\mathbf{C}$ is equipped with the usual metric.

Theorem 6.1. *An open subset Ω of $\mathbf{C}$ is connected if and only if each pair of points $z, w \in \Omega$ can be joined by a polygonal path whose sides are parallel to the axes.*

Proof. (Sketch.) The result holds (and is uninteresting) when Ω is the empty set. We therefore first assume Ω is connected and nonempty. Choose $z \in \Omega$. Consider the set S of points that can be reached from z by such a polygonal path. Then S is not empty, as $z \in S$. Also, S is open; if we can reach w, then we can also reach points in a ball about w. On the other hand, the set T of points we cannot reach from z is also open. By the definition of connectedness, Ω is not the union of disjoint open subsets unless one of them is empty. Thus, as S is nonempty, T must be empty. Hence $S = \Omega$ and the conclusion holds.

The proof of the converse statement is similar; we prove its contrapositive. Assume Ω is not connected and nonempty. Then $\Omega = A \cup B$, where A and B are open, $A \cap B = \varnothing$, but neither A nor B is empty. One then checks that no polygonal path in Ω connects points in A to points in B. Hence the contrapositive of the converse statement holds. $\qquad\square$

We pause to state the intuitive characterization of connected subsets of the real line. A subset of $\mathbf{R}$ (in the usual metric) is connected if and only if it is an interval. By convention, the word interval includes the entire real line, the empty set, and semi-infinite intervals. We leave the proof as an exercise, with the following hints. Given a bounded connected set S, let α be its greatest lower bound and let β be its least upper bound. We claim that S must be the interval between α and β, perhaps including one or both of these end points. To check this assertion, suppose $\alpha < x_0 < \beta$. If x_0 is not in S, then $\{x \in S : x < x_0\}$ and $\{x \in S : x > x_0\}$ are open nonempty subsets of S violating the definition of connectedness. The same idea works when S is unbounded below or above. The converse (an interval is connected) proceeds by writing the interval as $A \cup B$, where A and B are nonempty *closed* sets. (Their complements are open.) Choosing points in each and successively bisecting the interval between them creates two monotone sequences with a common limit, which must then be in both sets. Hence $A \cap B$ is nonempty.

▶ **Exercise 1.36.** Complete the proof that a subset of $\mathbf{R}$ is connected if and only if it is an interval.

Finally we mention one more concept, distinct from *connectedness*, but with a similar name. Roughly speaking, an open and connected subset S of $\mathbf{C}$ is called *simply connected* if it has no holes. Intuitively speaking, S has a hole if there is a closed curve in S which surrounds at least one point not in S. For example, the complement of a point is open and connected, but it is not simply connected. The set of z for which $1 < |z| < 2$ is open and connected but not simply connected. In

Chapter 8 we will define the *winding number* of a closed curve about a point. At that time we give a precise definition of simple connectivity.

Figure 1.6. A connected and simply connected set.

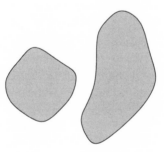

Figure 1.7. A disconnected set.

Figure 1.8. A connected but not simply connected set.

Chapter 2

Complex Numbers

The main point of our work in Chapter 1 was to provide a precise definition of the complex number field, based upon the existence of the real number field. While we will continue to work with the relationships between real numbers and complex numbers, our perspective will evolve toward thinking of complex numbers as the objects of interest. The reader will surely be delighted by how often this perspective leads to simpler computations, shorter proofs, and more elegant reasoning.

1. Complex conjugation

One of the most remarkable features of complex variable theory is the role played by complex conjugation. There are two square roots of -1, namely $\pm i$. When we make a choice of one of these, we create a kind of asymmetry. The mathematics must somehow keep track of the fundamental symmetry; these ideas lead to fascinating consequences.

Recall from Lemma 4.1 of Chapter 1 that $\mathbf{C}$ is not an ordered field. Therefore all inequalities used will compare real numbers. As we note below, real numbers are precisely those complex numbers unchanged by taking complex conjugates. Hence the fundamental issues involving inequalities also revolve around complex conjugation.

Definition 1.1. For x, y real, put $z = x + iy$. We write $x = \mathrm{Re}(z)$ and $y = \mathrm{Im}(z)$. The *complex conjugate* of z, written $\bar{z}$, is the complex number $x - iy$. The absolute value (or modulus) of z, written $|z|$, is the nonnegative real number $\sqrt{x^2 + y^2}$.

We make a few comments about the concepts in this definition. First we call x the *real part* and y the *imaginary part* of $x + iy$. Note that y is a real number; the imaginary part of z is not iy.

The absolute value function is fundamental in everything we do. Note first that $|z|^2 = z\bar{z}$. Next we naturally define the distance $\delta(z, w)$ between complex numbers z and w by $\delta(z, w) = |z - w|$. Then $\delta(z, w)$ equals the usual Euclidean distance between these points in the plane. We can use the absolute value function to define *bounded set*. A subset S of $\mathbf{C}$ is bounded if there is a real number M such that $|z| \leq M$ for all z in $\mathbf{C}$. Thus S is bounded if and only if S is a subset of a ball about 0 of sufficiently large radius.

The function mapping z into its complex conjugate is called *complex conjugation*. Applying this function twice gets us back where we started; that is, $\overline{\overline{z}} = z$. This function satisfies many basic properties; see Lemma 1.1.

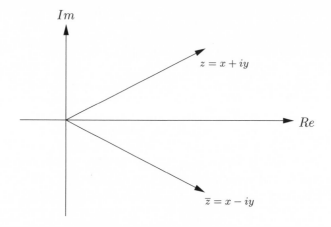

Figure 2.1. Complex conjugation.

Here is a way to stretch your imagination. Imagine that you have never heard of the real number system but that you know of a field called $\mathbf{C}$. Furthermore in this field there is a notion of convergent sequence making $\mathbf{C}$ complete in the sense of Cauchy sequences. Imagine also that there is a continuous function (called conjugation) $z \to \overline{z}$ satisfying properties 1), 2), and 3) from Lemma 1.1. Continuity guarantees that the conjugate of a limit of a sequence is the limit of the conjugates of the terms. We could then define the real numbers to be those complex numbers z for which $z = \overline{z}$.

For us the starting point was the real number system $\mathbf{R}$, and we constructed $\mathbf{C}$ from $\mathbf{R}$. We return to that setting.

Lemma 1.1. *The following formulas hold for all complex numbers z and w.*

1) $\overline{\overline{z}} = z$.

2) $\overline{z + w} = \overline{z} + \overline{w}$.

3) $\overline{zw} = \overline{z}\,\overline{w}$.

4) $|z|^2 = z\overline{z}$.

5) $\operatorname{Re}(z) = \frac{z + \overline{z}}{2}$.

6) $\operatorname{Im}(z) = \frac{z - \overline{z}}{2i}$.

7) *A complex number z is real if and only if $z = \overline{z}$.*

8) $|\overline{z}| = |z|$.

Proof. Left to the reader as an exercise. □

▶ **Exercise 2.1.** Prove Lemma 1.1. In each case, interpret the formula using Figure 2.1.

▶ **Exercise 2.2.** For a subset S of $\mathbf{C}$, define S^* by $z \in S^*$ if and only if $\overline{z} \in S$. Show that S^* is bounded if and only if S is bounded.

▶ **Exercise 2.3.** Show for all complex numbers z and w that
$$|z + w|^2 + |z - w|^2 = 2(|z|^2 + |w|^2).$$
Interpret geometrically.

▶ **Exercise 2.4.** Suppose that $|a| < 1$ and that $|z| \leq 1$. Prove that
$$\left| \frac{z - a}{1 - \overline{a}z} \right| \leq 1.$$
Comment: This fact is important in non-Euclidean geometry.

▶ **Exercise 2.5.** Let c be real, and let $a \in \mathbf{C}$. Describe geometrically the set of z for which $az + \overline{a}z = c$.

▶ **Exercise 2.6.** Let c be real, and let $a \in \mathbf{C}$. Suppose $|a|^2 \geq c$. Describe geometrically the set of z for which $|z|^2 + az + \overline{a}z + c = 0$.

▶ **Exercise 2.7.** Let a and b be nonzero complex numbers. Call them *parallel* if one is a *real* multiple of the other. Find a simple algebraic condition for a and b to be parallel. (Use the imaginary part of something.)

▶ **Exercise 2.8.** Let a and b be nonzero complex numbers. Find an algebraic condition for a and b to be perpendicular. (Use a similar idea as in Exercise 2.7.)

▶ **Exercise 2.9.** What is the most general (defining) equation for a line in $\mathbf{C}$? (Hint: The imaginary part of something must be 0.) What is the most general (defining) equation of a circle in $\mathbf{C}$?

▶ **Exercise 2.10.** For $z, w \in \mathbf{C}$, prove that $|\mathrm{Re}(z)| \leq |z|$ and $|z + w| \leq |z| + |w|$. Then verify that the function $\delta(z, w) = |z - w|$ defines a distance function making $\mathbf{C}$ into a metric space. (See Definition 6.1 of Chapter 1.)

2. Existence of square roots

In this section we give an algebraic proof that we can find a square root of an arbitrary complex number. Some subtle points arise in the choice of signs. Later we give an easier geometric method.

Proposition 2.1. *For each $w \in \mathbf{C}$, there is a $z \in \mathbf{C}$ with $z^2 = w$.*

Proof. Given $w = a + bi$ with a, b real, we want to find $z = x + iy$ such that $z^2 = w$. If $a = b = 0$, then we put $z = 0$. Hence we may assume that $a^2 + b^2 \neq 0$. The equation $(x + iy)^2 = w$ yields the system of equations $x^2 - y^2 = a$ and $2xy = b$. We convert this system into a pair of linear equations for x^2 and y^2 by writing

(1) $$(x^2 + y^2)^2 = (x^2 - y^2)^2 + 4x^2y^2 = a^2 + b^2.$$

The right-hand side of (1) is positive, and hence by Theorem 3.1 of Chapter 1 it has a positive real square root, and hence two real square roots. We choose the positive square root. We obtain the system

$$x^2 + y^2 = \sqrt{a^2 + b^2},$$
$$x^2 - y^2 = a.$$

We solve these two equations by adding and subtracting, obtaining

(2)
$$x^2 = \frac{a + \sqrt{a^2 + b^2}}{2},$$

(3)
$$y^2 = \frac{-a + \sqrt{a^2 + b^2}}{2}.$$

First note that the right-hand sides of (2) and (3) are nonnegative, because $a^2 \leq a^2 + b^2$, and hence $\pm a \leq \sqrt{a^2 + b^2}$. Recall that we chose the positive square root of the expression $a^2 + b^2$. Now we would like to take the square roots of the right-hand sides of (2) and (3) to define x and y, but we are left with some ambiguity of signs. In general there are two possible signs for x and two possible signs for y, leading to four candidates for the solution. Yet we know from Lemma 2.1 of Chapter 1 that only two of these can work.

We resolve this ambiguity in the following manner, consistent with our convention that $\sqrt{t}$ denotes the positive square root of t when $t > 0$. First we deal with the case $b = 0$. When $b = 0$, we put $y = 0$ if $a > 0$; we obtain the two solutions $\pm\sqrt{|a|}$. When $b = 0$, we put $x = 0$ if $a < 0$; we obtain the two solutions $\pm i\sqrt{|a|}$. In both of these cases we use $|a|$ for the square root of a^2.

Next suppose $b > 0$. In taking the square roots of (2) and (3), we choose x and y to have the same sign. Squaring now shows that these two answers satisfy $(x + iy)^2 = a + ib$. Finally suppose $b < 0$. In taking the square roots of (2) and (3), we choose x and y to have opposite signs. Squaring again shows that both answers satisfy $(x + iy)^2 = a + ib$. □

In the proof of Proposition 2.1, we obtained four candidates $\pm x \pm iy$ for z. When x and y are both not 0, these four candidates are distinct. As we noted in the proof, at most two of them can be square roots of w. Thus two of them fail. Hence the delicate analysis involving the signs is required. Things seem too complicated! On the other hand, the existence of square roots follows easily from the polar representation of complex numbers in Section 6. At that time we will develop geometric intuition clarifying the subtleties in the proof of Proposition 2.1.

Example 2.1. We solve $z^2 = 11 + 60i$ by the method of Proposition 2.1. We have $a^2 + b^2 = 121 + 3600$ and hence $\sqrt{a^2 + b^2} = 61$. Therefore $x^2 = \frac{11+61}{2} = 36$ and $y^2 = \frac{-11+61}{2} = 25$. We then take $x = 6$ and $y = 5$ or $x = -6$ and $y = -5$ to obtain the answers $z = \pm(6 + 5i)$. The other combinations of signs fail. If instead we want the square root of $11 - 60i$, then we have $x^2 = 36$ and $y^2 = 25$ as before, but we need to choose x and y to have opposite signs.

▶ **Exercise 2.11.** Find the error in the following alleged proof that $-1 = 1$.

$$-1 = i^2 = \sqrt{-1}\sqrt{-1} = \sqrt{(-1)(-1)} = \sqrt{1} = 1.$$

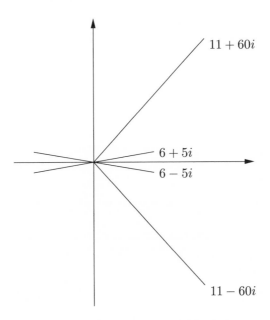

$11 + 60i$

$6 + 5i$

$6 - 5i$

$11 - 60i$

Figure 2.2. Finding square roots.

It is possible to solve cubic (third degree) and quartic (fourth degree) equations by using the quadratic formula cleverly. The history of this approach is quite interesting and relevant for the development of modern mathematics. See for example [**9**]. We limit ourselves here however to a few exercises about solving cubic polynomial equations. Cardano's solution of the cubic equation dates to 1545 and provided perhaps the first compelling argument in support of complex variables.

▶ **Exercise 2.12.** Let $z^3 + Az^2 + Bz + C$ be a cubic polynomial. What substitution reduces it to the form $w^3 + aw + b$?

▶ **Exercise 2.13.** Suppose we can solve the cubic $w^3 + aw + b = 0$, in the sense that we can find formulas for the roots in terms of a, b. By the previous exercise we can then solve the general cubic. To solve $w^3 + aw + b = 0$, we first make the substitution $w = \zeta + \frac{\alpha}{\zeta}$. If we choose α intelligently, then we get a sixth degree equation of the form

$$(4) \qquad \zeta^6 + c_3\zeta^3 + c_0 = 0.$$

What is the intelligent choice for α? Why? Since (4) is a quadratic in ζ^3, one can solve it by the quadratic formula.

▶ **Exercise 2.14.** Solve $z^3 + 3z - 4 = 0$ by the method of the previous exercise. Also solve it by elementary means and compare what you get. Do the same for $z^3 + 6z - 20 = 0$.

Remark 2.1. The method of Exercises 2.12 and 2.13 gives a formula for the solution of the general cubic equation $z^3 + Az^2 + Bz + C = 0$ in terms of A, B, C. Unfortunately the solution will involve nested radicals. Trying to simplify these nested radicals often leads one back to the original equation. Hence the method is somewhat unsatisfying. A similar discussion applies to fourth degree equations. Finally, no formula for the roots of the general polynomial equation of degree five and higher can possibly exist. See [**9**] for a beautiful treatment of both the mathematics and its history.

3. Limits

A sequence of complex numbers is a function from $\mathbf{N}$ (or from $\mathbf{N} \cup \{0\}$) to $\mathbf{C}$. We usually write the function by listing its values; we use the notation $\{z_n\}$ or $z_1, z_2, z_3, \ldots$ to denote a sequence of complex numbers. The crucial concept is the limit of a sequence.

Definition 3.1. Let $\{z_n\}$ be a sequence of complex numbers. Assume $L \in \mathbf{C}$. We say "the limit of z_n is L" or "z_n converges to L", and we write $\lim_{n\to\infty} z_n = L$ if the following statement holds:

   For all $\epsilon > 0$, there is an $N \in \mathbf{N}$ such that $n \geq N$ implies $|z_n - L| < \epsilon$.

We say that $\{z_n\}$ is a *Cauchy sequence* if the following statement holds:

   For all $\epsilon > 0$, there is an $N \in \mathbf{N}$ such that $m, n \geq N$ implies $|z_m - z_n| < \epsilon$.

We will not repeat standard facts about limits. The proofs are virtually the same as for limits of sequences of real numbers. We do make a few remarks and mention a few examples.

Remark 3.1. z_n converges to L if and only if $\mathrm{Re}(z_n)$ converges to $\mathrm{Re}(L)$ and $\mathrm{Im}(z_n)$ converges to $\mathrm{Im}(L)$. The proof is immediate from (5):

$$(5) \qquad |z_n - L|^2 = |\mathrm{Re}(z_n) - \mathrm{Re}(L)|^2 + |\mathrm{Im}(z_n) - \mathrm{Im}(L)|^2.$$

Remark 3.2. A sequence of complex numbers converges if and only if it is a Cauchy sequence; this fundamental fact follows from the previous remark and the corresponding property of the real number system.

▶ **Exercise 2.15.** Prove the two previous remarks about limits.

▶ **Exercise 2.16** (Continuity of conjugation). Assume z_n converges to L. Prove that $\overline{z_n}$ converges to $\overline{L}$.

The following lemma nicely illustrates some basic techniques in analysis and we apply it to study the geometric series.

Lemma 3.1. *For $|z| < 1$, we have $\lim_{N\to\infty} z^N = 0$. For $|z| > 1$, the sequence z^N diverges.*

Proof. First assume $|z| < 1$. By the definition of a limit (involving ϵ), or by Proposition 4.1 below, it suffices to show that $\lim_{N\to\infty} |z|^N = 0$. Let $t_n = |z|^n$; then (6) holds:

$$(6) \qquad\qquad t_{n+1} = |z| t_n.$$

Since $|z| < 1$, the sequence $\{t_n\}$ is monotone (nonincreasing). It is bounded below by 0. By Proposition 3.3 of Chapter 1, the limit exists. Call it L. Letting n tend to infinity in (6) gives $L = |z|L$. Since $|z| \neq 1$, the only possibility is $L = 0$.

If $|z| > 1$, then the sequence z^N is unbounded and hence diverges. $\qquad\qquad\square$

Example 3.1. How does the limiting behavior of z^n depend on z? By Lemma 3.1, $\lim_{N \to \infty} z^N = 0$ if $|z| < 1$. If $z = 1$, the limit equals 1. For other z with $|z| = 1$, the limit does not exist, as the sequence bounces around. If $|z| > 1$, the limit does not exist as a complex number; later we will allow infinity as a possible limit, and z^n does converge to infinity if $|z| > 1$.

Limits of function values rather than limits of sequences often arise. It is possible to reduce limits of function values to limits of sequences, but it is somewhat simpler to give the definition in terms of ϵ and δ. When we write $\lim_{z \to a} f(z) = L$, we ignore what happens at a itself. To deal with this situation, as in calculus, we say that $\lim_{z \to a} f(z) = L$ if for every $\epsilon > 0$ there is a $\delta > 0$ such that $0 < |z - a| < \delta$ implies $|f(z) - L| < \epsilon$. If we wanted to express this concept in terms of sequences, then we would need to consider all sequences tending to a for which z_n is never equal to a. By comparison, the definition of *continuous* (Definition 3.8 of Chapter 1) does take the point a into account, and therefore all sequences tending to a get considered.

Complex analytic functions. As in calculus, certain particular limits play a major role. The three most important limits are the definition of derivative, the definition of integral, and the definition of convergent infinite series. These three concepts are tied together in complex analysis even more closely than they are in calculus. To prepare for further developments, we make now a provisional definition of complex analytic function. In Chapter 6 we establish that the class of functions defined by this notion of complex differentiability is the same as the class of functions defined locally by convergent power series.

The notion of complex differentiability requires working on an open set. We defined *open set* in the metric space setting in Section 6 of Chapter 1. We formally define open set in **C** in Definition 1.1 of Chapter 3. For now, we need only the following intuitive idea: for f to be complex differentiable at p, it is necessary that f be defined at all points near p, in order that we can take the following limit.

Definition 3.2. Let Ω be an open subset of **C**, and suppose $f : \Omega \to \mathbf{C}$ is a function. We say that f is *complex analytic* on Ω if f is complex differentiable at each $p \in \Omega$. In other words f is complex analytic if at each p the following limit exists:

$$f'(p) = \lim_{\zeta \to 0} \frac{f(p + \zeta) - f(p)}{\zeta}.$$

The usual formal rules of calculus apply; for example, if $f(z) = z^n$ for a positive integer n, then f is complex analytic and $f'(z) = nz^{n-1}$. The derivative of a sum is the sum of the derivatives: $f + g$ is complex differentiable at z and $(f + g)'(z) = f'(z) + g'(z)$ if both $f'(z)$ and $g'(z)$ exist. The product rule for derivatives holds. Furthermore, if f is complex analytic and $f(z) \neq 0$, then $\frac{1}{f}$ is complex analytic near z, and its derivative is $\frac{-f'(z)}{f(z)^2}$. Therefore the quotient of two complex analytic functions is complex analytic at all points where the denominator does not vanish. Also, the chain rule holds. The proofs of these statements are essentially the same as in elementary calculus and they are given as exercises in Chapter 5. By the end of this book we will establish many beautiful results about complex analytic functions by combining the notions of derivatives, integrals, and convergent infinite series.

4. Convergent infinite series

We return to the basic properties of convergent infinite series. An infinite series arises when one tries to perform infinitely many additions. Let $\{z_n\}$ be a sequence in $\mathbf{C}$. We want to make sense out of the infinite sum $\sum_{n=0}^{\infty} z_n$. To do so, we first introduce the *partial sum* S_N defined by $S_N = \sum_{n=0}^{N} z_n$.

Definition 4.1. We say that $\sum_{n=0}^{\infty} z_n = L$ if $\lim_{N\to\infty} S_N = L$. We also then say that $\sum_{n=0}^{\infty} z_n$ *converges* to L. If the limit of the partial sums does not exist, then we say that $\sum_{n=0}^{\infty} z_n$ *diverges*.

The concept of infinite series closely parallels the writing of a real number as a decimal expansion. Some subtle ideas about power series in complex analysis naturally generalize simple things. These ideas are easier to grasp when we compare them with decimal expansions. See especially Section 6 of Chapter 4.

Let x be a real number with $0 \le x < 1$. We may express x as a decimal:

$$x = .a_1 a_2 a_3... = \sum_{n=1}^{\infty} a_n \left(\frac{1}{10}\right)^n.$$

In the decimal expansion we demand that each coefficient a_n be an integer between 0 and 9. The convergence of the series follows from the following reasoning. Since each term is nonnegative, the sequence S_N of partial sums of the expansion is nondecreasing. By Proposition 3.3 and Remark 3.2 (both from Chapter 1), such a sequence has a limit if and only if it is bounded. Note that $S_N \le \sum_{n=1}^{N} 9(\frac{1}{10})^n = 1 - \frac{1}{10^N} \le 1$ for all N. Therefore S_N is bounded and the decimal expansion makes sense.

To make the previous paragraph more precise and to prepare for later discussion, we pause to mention the geometric series. We discuss it in detail in Chapter 4. The distributive law and induction yield

$$(1-z) \sum_{n=0}^{N-1} z^n = 1 - z^N,$$

and therefore for $z \ne 1$ we have

$$(7) \qquad \sum_{n=0}^{N-1} z^n = \frac{1 - z^N}{1 - z}.$$

We call (7) the finite geometric series.

Next assume that $|z| < 1$. By Lemma 3.1, $\lim_{N\to\infty} z^N = 0$. Combining this limit with the finite geometric series gives the *geometric series*:

$$(8) \qquad \sum_{n=0}^{\infty} z^n = \lim_{N\to\infty} \sum_{n=0}^{N-1} z^n = \lim_{N\to\infty} \frac{1 - z^N}{1 - z} = \frac{1}{1 - z}$$

whenever $|z| < 1$. The geometric series appeared implicitly in the proof of the decimal expansion when we noted that the decimal $.999...99$ (a total of N nines) equals

$1 - (\frac{1}{10})^N$. Definition 4.1 implies that (the infinite decimal expansion) .999... = 1; there can be no debate about this matter, despite innumerable internet discussions about it!

Also by Lemma 3.1, the limit as N tends to infinity of (7) does not exist if $|z| > 1$. If $|z| = 1$ but $z \neq 1$, the limit also fails to exist. Since $z = 1$ is precluded from the discussion, we conclude that the geometric series converges if and only if $|z| < 1$.

The next several exercises follow immediately from Definition 4.1. The second exercise provides a distributive law for convergent infinite sums; this law gets used in one of the standard proofs that .999... = 1.

▶ **Exercise 2.17.** Assume that both $\sum_{n=0}^{\infty} a_n$ and $\sum_{n=0}^{\infty} b_n$ converge. Prove that $\sum_{n=0}^{\infty}(a_n + b_n)$ converges.

▶ **Exercise 2.18** (Infinite distributive law). Assume that $\sum_{n=0}^{\infty} a_n$ converges and $c \in \mathbf{C}$. Prove that $c \sum_{n=0}^{\infty} a_n = \sum_{n=0}^{\infty} c a_n$.

▶ **Exercise 2.19.** Fill in the details of the following proof that .999... = 1. First regard .999... as an infinite series. Show using Proposition 3.3 of Chapter 1 that it converges to something, say L. Then use the infinite distributive law (Exercise 2.18) to conclude that $10\, L = 9 + L$. Hence $L = 1$.

▶ **Exercise 2.20.** Prove by induction that $|\sum_1^n c_j| \leq \sum_1^n |c_j|$.

▶ **Exercise 2.21.** Consider a sports tournament with 64 teams. In the first round all teams play. The 32 winners advance to the second round, and so on, until there is one winner. Show **in two ways** that there are 63 games played in total. What does this argument have to do with the finite geometric series?

The most fundamental tests for convergence are *comparison tests*. The term *comparison test* is standard from calculus courses that provide recipes for deciding the convergence of infinite series. We prove this test next, and we include with it the simple version for sequences.

Proposition 4.1. 1) *Assume $c_n \geq 0$ for all n and that $\lim(c_n) = 0$. If $z_n \in \mathbf{C}$ and $|z_n| \leq c_n$ for all n, then $\lim(z_n) = 0$ also.*

2) *Assume $b_n \geq 0$ for all n and that $\sum_1^{\infty} b_n$ converges. If $a_n \in \mathbf{C}$ and $|a_n| \leq b_n$ for all n, then $\sum_1^{\infty} a_n$ also converges.*

Proof. 1) Given $\epsilon > 0$, we can choose an N such that $|c_n - 0| < \epsilon$ whenever $n \geq N$. For the same N and $n \geq N$ we obtain

$$|z_n - 0| \leq c_n = |c_n - 0| < \epsilon.$$

2) This proof displays the power of the Cauchy criterion for convergence. Let T_N denote the N-th partial sum of the series $\sum_1^{\infty} b_n$, and let S_N denote the N-th partial sum of the series $\sum_1^{\infty} a_n$. To show that the sequence S_N converges, we show that it is a Cauchy sequence. For $M > N$ we have

$$(9) \qquad |S_M - S_N| = |\sum_{n=N+1}^{M} a_n| \leq \sum_{n=N+1}^{M} |a_n| \leq \sum_{n=N+1}^{M} b_n = |T_M - T_N|.$$

The sequence T_N converges and hence is Cauchy; thus we can make the right-hand side of (9) as small as we wish by choosing N large enough. By (9), S_N is also Cauchy and therefore converges. $\square$

We next recall from calculus one form of a test for convergence of a series called the *ratio test*. In both the ratio test and the root test, the idea behind the test is to find a quantity that plays the role r does in the geometric series $\sum r^n$.

Proposition 4.2. *Assume $a_n \neq 0$ for all n. Suppose the limit $L = \lim_{n\to\infty} \frac{|a_{n+1}|}{|a_n|}$ of the ratios of the absolute values of successive terms exists and that $L < 1$. Then the series $\sum_{n=0}^{\infty} a_n$ converges. If $L > 1$, then the series diverges. The test is inconclusive when $L = 1$.*

Proof. We sketch the convergence part; the divergence part is similar and easier and hence is left to the reader as an exercise. The intuition behind both parts is the comparison with the geometric series. Let $b_n = |a_n|$. By the comparison test, to show convergence, it suffices to show that $\sum b_n$ converges when $L < 1$. For large enough n, $\frac{b_{n+1}}{b_n}$ is approximately L, which we assume is less than 1. Then there is a number r with $0 < r < 1$ and $\frac{b_{n+1}}{b_n} < r$ for all sufficiently large n, say $n \geq N$. One can then compare the tail $\sum_{n \geq N} b_n$ with the convergent geometric series $b_N r^N \sum_{j=0}^{\infty} r^j$. Thus the series converges when $L < 1$. $\square$

▶ **Exercise 2.22.** Prove the divergence part of the ratio test. Give an example where $L = 1$ and the series converges. Give an example where $L = 1$ and the series diverges. Formulate and prove the *root test*; to do so, consider $L = \lim \sup |a_n|^{\frac{1}{n}}$.

More on infinite series. We make a few comments about absolute convergence. Let $\{a_n\}$ be a sequence of complex numbers. We say that $\sum_{n=1}^{\infty} a_n$ *converges absolutely* if $\sum_{n=1}^{\infty} |a_n|$ converges. When $\sum_{n=1}^{\infty} a_n$ converges but $\sum_{n=1}^{\infty} |a_n|$ does not, we say that $\sum_{n=1}^{\infty} a_n$ *converges conditionally*. Absolute convergence is easier to understand than is conditional convergence. The next several exercises illustrate this point rather well. It is particularly striking that the limit of a conditionally convergent series depends on the order in which we add up the terms. We use the term *rearrangement* for the series obtained by performing the sum in a different order.

▶ **Exercise 2.23.** Show that $\sum_{n=1}^{\infty} \frac{(-1)^{n+1}}{n}$ converges conditionally. The value of the sum is $\log(2)$, approximately $.69$. Show that by grouping the terms by taking two positive terms, then one negative term, then two positive terms, then one negative term, and so on, the series adds up to a number larger than 1. The value of the sum therefore depends on the order in which the terms are summed.

▶ **Exercise 2.24.** Express the series described in the previous exercise (after grouping) in the form $\sum_{n=1}^{\infty} b_n$, and find a simple expression for b_n. Try to find this infinite sum.

▶ **Exercise 2.25** (Riemann's remark). Assume that $a_n \in \mathbf{R}$ and that $\sum_{n=1}^{\infty} a_n$ converges conditionally. Fix an arbitrary $L \in \mathbf{R}$. Show that we can rearrange the terms to make the sum converge to L.

▶ **Exercise 2.26** (Complex variable analogue). (Difficult) Assume that $a_n \in \mathbf{C}$ and consider rearrangements of the convergent series $\sum_{n=1}^{\infty} a_n$. Show that each of the following situations is possible and that this list includes all possibilities.

1) $\sum_{n=1}^{\infty} a_n$ converges absolutely and hence all rearrangements converge to the same value.

2) The set of possible values of convergent rearrangements is all of $\mathbf{C}$.

3) The set of possible values of convergent rearrangements is an arbitrary line in $\mathbf{C}$.

The book [**15**] is devoted almost entirely to the kind of question from Exercise 2.26 in various settings.

▶ **Exercise 2.27.** Find a (necessarily conditionally convergent) series (of real or complex numbers) such that $\sum a_n$ converges but $\sum a_n^3$ diverges.

5. Uniform convergence and consequences

This section describes some important results in basic analysis. The main point is Theorem 5.3, which is used to justify differentiating or integrating a power series term by term. Many readers will wish to skip this section and return to it when needed.

The examples we have seen indicate some of the subtleties arising when absolute convergence fails. For series of functions we need a second way to improve the notion of convergence. Note that the power series $\sum_1^{\infty} a_n z^n$ can be regarded as summing the sequence of functions f_n, where $f_n(z) = a_n z^n$. Given a sequence or series of functions, things work best when the sequence or series *converges uniformly*, a concept defined below. As in the relationship between sequences and series of numbers, it suffices to study various types of convergence for sequences of functions and to apply them to sequences of partial sums when we work with series of functions.

Let f_n be a sequence of real- or complex-valued functions defined on some set E. We say that f_n *converges pointwise* on E if, for each $z \in E$, $f_n(z)$ converges. This notion is inadequate for many purposes. We introduce the crucial stronger condition of uniform convergence.

Definition 5.1. Let E be a subset of $\mathbf{C}$. Let $f_n : E \to \mathbf{C}$ be a sequence of functions defined on E. We say that f_n converges uniformly to f on E if, for all $\epsilon > 0$, there is an N such that $n \geq N$ implies $|f_n(z) - f(z)| < \epsilon$ for all $z \in E$. We say that an infinite series converges uniformly if its sequence of partial sums converges uniformly.

The difference between pointwise convergence and uniform convergence arises from the location of the quantifier "for all z". In Definition 5.1, the N chosen must work for all z in E. In the definition of pointwise convergence, for each z there is an N, but there might be no single N that works for all z at the same time.

▶ **Exercise 2.28.** Discuss the convergence and uniform convergence of the given sequence of functions on the given set E.

1) Put $f_n(z) = z^n$. First let E be the real interval $[0, 1]$. Then let $E = \{z : |z| \leq R < 1\}$. Then let $E = \{z : |z| < 1\}$. What happens on the closed unit disk?

2) Put $f_n(z) = z^n(1 - z)$. Let E be the real interval $[0, 1]$.

3) Put $f_n(z) = nz^n(1 - z)$. Let E be the real interval $[0, 1]$.

4) Put $f_n(z) = z^n(1 - |z|)$. Let $E = \{z : |z| < 1\}$.

5) Let f_n be a sequence of functions on a set E. Suppose for each n we have $|f_n(z)| \leq M_n$ on E and assume $\sum M_n$ converges. Show that $\sum_{n=1}^{\infty} f_n(z)$ converges uniformly on E.

▶ **Exercise 2.29.** Give several distinct examples of a sequence of (real- or complex-valued) functions that converges pointwise on **R** but not uniformly. (Pictures might be better than formulas.)

We continue this section by discussing some technical real analysis and advanced calculus. Readers who wish to jump ahead to the next section and return to these results later are welcome to do so. Theorem 5.3 and its Corollary 5.1 play an important role in the rigorous development of complex analysis, but they are not necessary for many of our results. The theorem provides a circumstance in which the derivative of a limit of a sequence of functions is the limit of the derivatives. Its corollary allows us to differentiate or integrate a power series term by term within its circle of convergence.

Before we discuss these results, we provide many examples which reveal why there is something to prove. Integrals and derivatives are themselves limits. One cannot interchange the order of limits in general, as the examples illustrate.

A function of one or several variables is called *smooth* if its partial derivatives of all orders exist and are themselves continuous functions. The word smooth will also be used in Chapter 6 to describe certain curves. In the results from calculus used in this book, only continuous derivatives of the first order will matter. Hence we often use the following standard but awkward term. A function of one or several real variables is called *continuously differentiable* if its first partial derivatives exist at each point and are themselves continuous functions. This assumption arises at various places in our development, although for complex analytic functions, the assumption is not needed. See Corollary 4.3 of Chapter 6.

We first give an example of a differentiable function f on the real line whose derivative is not continuous; thus f is not continuously differentiable. We continue with examples illustrating the subtleties that can arise when we interchange the order of limiting operations.

Example 5.1. For $x \neq 0$, put $f(x) = x^2\sin(\frac{1}{x})$; put $f(0) = 0$. We claim that $f'(x)$ exists for all x. The product rule for derivatives gives for $x \neq 0$

$$f'(x) = 2x\sin(\frac{1}{x}) - \cos(\frac{1}{x}),$$

and the definition of $f'(0)$ as a limit quotient gives $f'(0) = 0$. Thus the function f' makes sense. On the other hand, $\lim_{x \to 0} f'(x)$ does not exist, and therefore f' is not continuous at 0.

Example 5.2. Put $f_n(x) = n$. Then $f_n'(x) = 0$ for all x, but $\lim f_n(x)$ does not exist. The limit of the derivatives is not the derivative of the limit.

Example 5.3. Put $f_n(x) = \frac{x^{n+1}}{n+1}$. For all $x \in [-1, 1]$ we have $\lim f_n(x) = 0$. Also, $f_n'(x) = x^n$ for all x. The limit of $f_n'(x)$ is zero if $|x| < 1$, but this limit is 1 if $x = 1$ and it does not exist if $x = -1$. Again the limit of the derivatives is not the derivative of the limit.

Example 5.4. Put $f_n(x) = n$ for $0 < x \leq \frac{1}{n}$ and $f_n(x) = 0$ otherwise. Then $\int_0^1 f_n(x)dx = 1$ for all n. Also $\lim f_n(x) = 0$ for all x. The limit of the integral is not the integral of the limit.

In each of the previous three examples, a difficulty arises because two limits are being taken. The order in which they are taken matters. The following shockingly simple example is worth some thought.

Example 5.5. Put $f(x, y) = |x|^{|y|}$. Then $\lim_{x \to 0} \lim_{y \to 0} f(x, y) = 1$, whereas $\lim_{y \to 0} \lim_{x \to 0} f(x, y) = 0$. The limit changes when we reverse the order in which the variables go to zero. In particular there is no value of $f(0, 0)$ making f continuous there.

It should now be evident that we need to find hypotheses enabling us to interchange the order of limiting operations. The following results from analysis can be regarded as filling that need.

Lemma 5.1. *Suppose $f_n : E \to \mathbf{C}$ is continuous and f_n converges uniformly on E to some function f. Then f is continuous on E.*

Proof. The key point is an $\frac{\epsilon}{3}$ argument. By the triangle inequality we have, for all $z, w \in E$ and for all n,

$$(10) \qquad |f(z) - f(w)| \leq |f(z) - f_n(z)| + |f_n(z) - f_n(w)| + |f_n(w) - f(w)|.$$

We can make the first and third terms on the right-hand side as small as we wish by making n large. We can make the middle term as small as we wish by making z close to w. We prove continuity at z by the ϵ-δ definition. Given $\epsilon > 0$, we choose N such that $n \geq N$ implies $|f_n(\zeta) - f_n(w)| < \frac{\epsilon}{3}$ for all ζ and w. Such an N exists by uniform convergence. Fix an n with $n \geq N$. We choose $\delta > 0$ small enough to guarantee that $|f_n(z) - f_n(w)| < \frac{\epsilon}{3}$ when $|z - w| < \delta$. Such a δ exists because f_n is continuous at z. Then (10) implies that $|f(z) - f(w)| < \epsilon$. $\qquad \square$

In the next two results we assume that $E = [a, b]$ is a bounded closed interval on $\mathbf{R}$. In both proofs we use the inequality $|\int g| \leq \int |g|$; this simple fact is discussed in detail in Section 1 of Chapter 6.

Lemma 5.2. *Suppose $f_n : E \to \mathbf{C}$ is continuous for each n and that f_n converges uniformly to f on E. Then $\int_a^b f_n$ converges to $\int_a^b f$.*

Proof. If $a = b$, there is nothing to prove, so assume that $a < b$. By Lemma 5.1, f is continuous on $[a, b]$ and hence integrable. Given $\epsilon > 0$, by uniform convergence we can find an N such that $n \geq N$ implies, for all $t \in [a, b]$,

$$|f_n(t) - f(t)| \leq \frac{\epsilon}{b - a}.$$

To show that $\int_a^b f_n$ converges to $\int_a^b f$, we simply note for $n \geq N$ that

$$\left| \int_a^b f_n(t)dt - \int_a^b f(t)dt \right| \leq \int_a^b |f_n(t) - f(t)|dt \leq (b-a)\frac{\epsilon}{b-a} = \epsilon.$$

We have verified the definition of limit for sequences.   □

Theorem 5.1. *Suppose $f_n : E \to \mathbf{C}$ is a sequence of continuously differentiable functions. Thus f_n and $\frac{df_n}{dx}$ are continuous on $[a, b]$. Assume that the sequence of numbers $f_n(a)$ converges. Finally assume that the sequence $\frac{df_n}{dx}$ converges uniformly to some function g. Then f_n converges uniformly on E to a function f, the function f is continuously differentiable, and $\frac{df}{dx} = g$. In other words,*

$$\frac{d}{dx}\lim(f_n) = \lim\frac{df_n}{dx}.$$

Proof. We require the fundamental theorem of calculus. Using it, we write

$$(11) \qquad\qquad f_n(x) = \int_a^x \frac{df_n}{dx}(t)dt + f_n(a).$$

Then we let n tend to infinity in (11). The first term on the right-hand side of (11) converges by Lemma 5.2 to $\int_a^x g(t)dt$ and the second term converges to some number. Therefore, for each x, $f_n(x)$ converges to some number we call $f(x)$. Below we check that the convergence is uniform. First note that

$$(12) \qquad\qquad f(x) = \int_a^x g(t)dt + f(a).$$

The function g is continuous by Lemma 5.1. The fundamental theorem of calculus therefore applies, and we obtain $\frac{df}{dx} = g$. Since g is continuous, f is continuously differentiable. It remains to show that f_n converges to f uniformly. To do so, subtract (12) from (11) to obtain a formula for $f_n(x) - f(x)$. Taking absolute values and using an $\frac{\epsilon}{2}$ argument shows that it suffices to estimate

$$\left| \int_a^x \left(\frac{df_n}{dx}(t) - g(t) \right) dt \right|$$

uniformly in x. By using $|\int h| \leq \int |h|$, it suffices to show that $\frac{df_n}{dx}$ converges uniformly to g, which is one of the assumptions.   □

We now turn to functions of several variables. Although the next definition and theorem make sense in higher dimensions, we restrict our consideration to two dimensions. Versions of the next definition and theorem appear in many calculus books. See for example Chapter 12 of [**24**].

Definition 5.2. Let Ω be an open subset of $\mathbf{R}^2$ and suppose $f : \Omega \to \mathbf{C}$. We say that f is *differentiable* at (x, y) if there are complex numbers a, b such that

$$(13) \qquad\qquad f(x + h, y + k) = f(x, y) + ah + bk + E(h, k),$$

where

$$(14) \qquad\qquad \lim_{(h,k)\to(0,0)} \frac{E(h, k)}{\sqrt{h^2 + k^2}} = 0.$$

This definition states that f is approximately linear at (x, y). The expression $E(h, k)$ in (14) is regarded as an error term, and the existence of the limit means that $E(h, k)$ tends to 0 faster than the length of the vector (h, k). The derivative, or gradient vector $\nabla f(x, y) = (a, b)$, provides the linear approximation: for (h, k) small, $f(x+h, y+k)$ equals approximately $f(x, y)$ plus the linear correction $ah + bk$. When f is differentiable at (x, y), we have

$$\nabla f(x, y) = (\frac{\partial f}{\partial x}(x, y), \frac{\partial f}{\partial y}(x, y)).$$

The existence of these partial derivatives does not imply that f is differentiable. If these partial derivatives are *continuous*, then the following important result from advanced calculus does guarantee differentiability.

Theorem 5.2. *Let Ω be an open subset of $\mathbf{R}^2$ and suppose $f : \Omega \to \mathbf{C}$. Suppose that the partial derivatives $\frac{\partial f}{\partial x}$ and $\frac{\partial f}{\partial y}$ exist and are continuous on Ω. Then f is differentiable.*

Proof. The idea of the proof is quite simple. We use the mean value theorem from basic calculus. If g is a differentiable function of one variable and the interval (x_0, x_1) lies in the domain of g, then there is a c (depending on x_1 and x_2) in this interval such that

$$(15) \qquad g(x_2) - g(x_1) = g'(c)(x_2 - x_1).$$

Of course, if g is continuously differentiable, then we may let x_2 tend to x_1 and then $g'(c)$ tends to $g'(x_1)$.

Let f satisfy the hypotheses of the theorem. Let $\nabla f(x, y) = (a, b)$. We study $f(x + h, y + k) - f(x, y)$ by reducing to the one-dimensional case as follows. Since Ω is open, there is an open rectangle about (x, y) lying within Ω. For (h, k) small we assume that $(x + h, y + k)$ is in this rectangle. We then write

$$(16) \quad f(x+h, y+k) - f(x, y) = f(x+h, y+k) - f(x+h, y) + f(x+h, y) - f(x, y).$$

The first two terms on the right-hand side of (16) involve only a change in the vertical direction, and the last two terms involve only a change in the horizontal direction. We use the mean value theorem on each of these pairs to get numbers c and δ such that c is between x and $x + h$, δ is between y and $y + k$, and

$$(17) \qquad f(x + h, y + k) - f(x, y) = \frac{\partial f}{\partial y}(x + h, \delta)k + \frac{\partial f}{\partial x}(c, y)h.$$

Now subtract $\nabla f(x, y) \cdot (h, k)$ from both sides of (17) and rearrange terms to get

$$f(x + h, y + k) - f(x, y) - \nabla f(x, y) \cdot (h, k)$$

$$(18) \qquad = \left(\frac{\partial f}{\partial y}(x + h, \delta) - \frac{\partial f}{\partial y}(x, y) \right) k + \left(\frac{\partial f}{\partial x}(c, y) - \frac{\partial f}{\partial x}(c, y) \right) h.$$

Since the partials are continuous at (x, y), the expressions multiplying h and k in (18) go to zero as (h, k) tends to $(0, 0)$. The factors of h and k also do of course. It follows that the right-hand side of (18) goes to zero faster than $\sqrt{h^2 + k^2}$, and f is differentiable. $\qquad \square$

Now that we have sketched an entire basic analysis course, we can derive some important consequences.

Theorem 5.3. *Let Ω be an open set in $\mathbf{C}$ and suppose that $F_N : \Omega \to \mathbf{C}$ is a sequence of continuously differentiable functions. Assume that F_N converges pointwise to some function F on Ω. Suppose furthermore that the partial derivatives $\frac{\partial F_N}{\partial x}$ and $\frac{\partial F_N}{\partial y}$ converge uniformly on each closed and bounded subset of Ω. Then the limit function F is differentiable at each point of Ω and its partial derivatives are continuous. These partials are given by*

$$(19) \qquad \frac{\partial F}{\partial x} = \lim_{N \to \infty} \frac{\partial F_N}{\partial x},$$

$$(20) \qquad \frac{\partial F}{\partial y} = \lim_{N \to \infty} \frac{\partial F_N}{\partial y}.$$

Proof. The result follows by combining Theorems 5.1 and 5.2. $\qquad\square$

A corollary of this theorem will be crucial in our development, and we therefore state it now in anticipation. We naturally want to differentiate a power series $\sum c_n z^n$, within its region of convergence, with respect to z. Recall that the power series is the limit of its partial sums. Under the assumption of uniform convergence of the derivatives with respect to x and y, Theorem 5.3 allows for the interchange of derivatives and infinite sum (limit of the partial sums). We can express the operations $\frac{\partial}{\partial z}$ and $\frac{\partial}{\partial \overline{z}}$ in terms of $\frac{\partial}{\partial x}$ and $\frac{\partial}{\partial y}$ and vice versa. See Definition 2.1 of Chapter 5. But a power series in z is somehow independent of $\overline{z}$, and we can ignore the $\frac{\partial}{\partial \overline{z}}$ derivatives. See Chapters 4, 5, and 6 for the full story.

Corollary 5.1. *Assume that $\sum_{n=0}^{\infty} c_n z^n$ converges on the set $\{|z| < R\}$, and call the sum $f(z)$. Then f is complex differentiable on $\{|z| < R\}$ and $f'(z) = \sum_{n=0}^{\infty} n c_n z^{n-1}$. Furthermore, the series $\sum_{n=0}^{\infty} \frac{c_n}{n+1} z^{n+1}$ also converges for $|z| < R$. It represents a complex differentiable function F for which $F'(z) = f(z)$.*

Proof. (Sketch) Assume that a power series converges on $\{|z| < R\}$. Its partial sums are polynomials in z and hence complex differentiable functions. Theorem 2.1 of Chapter 4 guarantees the following precise statement on convergence: for any r such that $r < R$, the series converges absolutely and uniformly for $\{|z| \le r\}$. We may therefore invoke the theorems of this section to interchange both differentiation and integration with infinite sum in $\{|z| \le r\}$. Since r is an arbitrary number smaller than R, the results are valid in the open disk $\{|z| < R\}$. $\qquad\square$

6. The unit circle and trigonometry

The unit circle S^1 is the set of complex numbers z at distance 1 from the origin. Thus $S^1 = \{z : |z| = 1\}$. Consider a point z on S^1. Assuming we know what cosine and sine are, we can also represent z by $z = \cos(\theta) + i\sin(\theta)$, where θ is the usual polar angle. The equation $\cos^2(\theta) + \sin^2(\theta) = 1$ is simply a way of rewriting the equation $|z|^2 = 1$. Complex variables provide an elegant approach to trigonometry; this beautiful viewpoint enables routine, unified proofs and it makes one wonder why high school students are subjected to so many complicated trig identities.

Rather than presuming any knowledge of trigonometry, we will start by defining the exponential function and go on to define sine and cosine in terms of it. This approach might be counterintuitive to some readers. We therefore also recall from calculus the power series expansions for cosine and sine as a way of justifying what we have done. See also Exercise 2.57 for a justification using differential equations. The main reason for developing trigonometry in this manner is that many complicated expressions in trigonometry simplify when they are expressed in terms of complex exponentials.

Our formal development begins with the exponential function, written $\exp(z)$ or e^z. We define this function by its power series

$$(21) \qquad e^z = \sum_{n=0}^{\infty} \frac{z^n}{n!}.$$

This series converges for all z by the ratio test: for $z \neq 0$ the absolute value of the ratio of successive terms is $\frac{|z|}{n+1}$, which tends to 0 as n tends to infinity. Since $0 < 1$, the ratio test guarantees convergence. For any closed disc $\{z : |z| \leq R\}$, the power series for e^z converges absolutely and uniformly on this closed disk. (See Theorem 2.1 of Chapter 4 for a general statement.) Corollary 5.1 then implies that the exponential function is complex differentiable. Furthermore, by differentiating term by term, we see that $\frac{d}{dz} e^z = e^z$. For now we do not use this information.

We note that $e^0 = 1$ and soon we will verify the fundamental identity $e^{z+w} = e^z e^w$. Something amazing then happens when we approach trigonometry via complex geometry. The trigonometric functions are defined in terms of the exponential function and all trigonometric identities become consequences of the functional equation $e^{z+w} = e^z e^w$ and the (almost obvious) identity (23) below.

Theorem 6.1. *For all $z, w \in \mathbf{C}$, we have $e^{z+w} = e^z e^w$.*

Proof. The series defining the exponential function converges absolutely for all z in $\mathbf{C}$, and hence the value of the sum does not depend upon the order of summation. The definition of the exponential function and the binomial theorem yield

$$(22) \qquad e^{z+w} = \sum_{n=0}^{\infty} \frac{(z+w)^n}{n!} = \sum_{n=0}^{\infty} \frac{1}{n!} \sum_{k=0}^{n} \binom{n}{k} z^k w^{n-k} = \sum_{n=0}^{\infty} \sum_{k=0}^{n} \frac{z^k}{k!} \frac{w^{n-k}}{(n-k)!}.$$

Now set $n = j + k$ in (22) and redetermine the limits of summation to get

$$e^{z+w} = \sum_{j=0}^{\infty} \sum_{k=0}^{\infty} \frac{z^k}{k!} \frac{w^j}{j!} = e^z e^w.$$

$\square$

The functional equation implies

$$e^z e^{-z} = e^0 = 1;$$

hence $e^z \neq 0$ and $\frac{1}{e^z} = e^{-z}$. Note also (using the continuity of conjugation) that

$$(23) \qquad e^{\bar{z}} = \overline{e^z}.$$

Next we consider, for t real, the expression e^{it}. By (23) its conjugate is e^{-it}, which is also its reciprocal. Hence $|e^{it}| = 1$, and e^{it} lies on the unit circle.

We make the link with trigonometry as follows. We define cosine and sine by the formulas

$$\text{(24)} \qquad \cos(z) = \frac{e^{iz} + e^{-iz}}{2},$$

$$\text{(25)} \qquad \sin(z) = \frac{e^{iz} - e^{-iz}}{2i}.$$

When z is real, (24) expresses $\cos(z)$ as the real part of e^{iz} and (25) expresses $\sin(z)$ as the imaginary part of e^{iz}. We emphasize that formulas (24) and (25) hold for all complex z.

▶ **Exercise 2.30.** Prove (23). Where is the continuity of conjugation used?

▶ **Exercise 2.31.** Show that $\cos^2(z) + \sin^2(z) = 1$.

We discuss further why formulas (24) and (25) are plausible definitions. Assume that we have defined cosine and sine in some other way and we know the following things: $\sin(0) = 0$, the derivative of sine is cosine, and the derivative of cosine is minus sine. By Taylor's formula and an estimate on the remainder we obtain the power series expansions for cosine and sine:

$$\text{(26)} \qquad \cos(t) = \sum_{j=0}^{\infty} \frac{(-1)^j t^{2j}}{(2j)!},$$

$$\text{(27)} \qquad \sin(t) = \sum_{j=0}^{\infty} \frac{(-1)^j t^{2j+1}}{(2j+1)!}.$$

On the other hand, for t real, consider the power series expansion of e^{it}. We obtain

$$
\begin{aligned}
e^{it} &= \sum_{n=0}^{\infty} \frac{(it)^n}{n!} = \sum_{j=0}^{\infty} \frac{(it)^{2j}}{(2j)!} + \sum_{k=0}^{\infty} \frac{(it)^{2k+1}}{(2k+1)!} \\
\text{(28)} \qquad &= \sum_{j=0}^{\infty} \frac{(-1)^j t^{2j}}{(2j)!} + i \sum_{k=0}^{\infty} \frac{(-1)^k t^{2k+1}}{(2k+1)!} = \cos(t) + i\sin(t).
\end{aligned}
$$

For real t, we therefore have $\cos(t) = \text{Re}(e^{it})$ and $\sin(t) = \text{Im}(e^{it})$. Thus (24) and (25) agree with what we expect when z is real; these definitions are more general, because cosine and sine are functions defined on all of $\mathbf{C}$. We define the other trigonometric functions in terms of cosine and sine in the usual manner; for example $\tan(z) = \frac{\sin(z)}{\cos(z)}$.

The following consequence of the functional equation (Theorem 6.1), dating back to at least 1707, is useful for proving trig identities.

Proposition 6.1 (De Moivre's formula). *For each positive integer n and each real number θ, we have*

$$\text{(29)} \qquad (\cos(\theta) + i\sin(\theta))^n = \cos(n\theta) + i\sin(n\theta).$$

Proof. Since $\cos(\phi) + i\sin(\phi) = e^{i\phi}$, (29) can be restated as $(e^{i\theta})^n = e^{in\theta}$. This restated version follows by induction: the result is trivial for $n = 1$ and the induction step follows from $e^{z+w} = e^z e^w$. $\qquad \square$

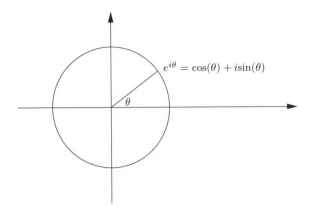

Figure 2.3. The unit circle.

Using Proposition 6.1, we derive some identities. Put $n = 2$ and $n = 3$ in de Moivre's formula and equate real and imaginary parts. We obtain

$$\cos(2\theta) = \cos^2(\theta) - \sin^2(\theta),$$

$$\sin(2\theta) = 2\sin(\theta)\,\cos(\theta),$$

$$\cos(3\theta) = \cos^3(\theta) - 3\cos(\theta)\sin^2(\theta),$$

$$\sin(3\theta) = 3\cos^2(\theta)\,\sin(\theta) - \sin^3(\theta).$$

▶ **Exercise 2.32.** Find formulas for $\cos(4\theta)$ and $\sin(4\theta)$.

▶ **Exercise 2.33.** Simplify the expressions $\cos(z + w) \pm \cos(z - w)$.

▶ **Exercise 2.34.** Express $\cos^2(z)$ and $\sin^2(z)$ in terms of $\cos(2z)$.

▶ **Exercise 2.35.** Express $(1 + i)^{10}$ in the form $x + iy$. (Do not expand!)

▶ **Exercise 2.36.** For $t \in \mathbf{R}$, show that the point $\left(\frac{1-t^2}{1+t^2}, \frac{2t}{1+t^2}\right)$ is on the unit circle. Express this point in the form $e^{i\theta}$. What is the relationship between t and θ?

For later use we define the hyperbolic functions cosh, sinh, and tanh by

$$\cosh(z) = \frac{e^z + e^{-z}}{2},$$

$$\sinh(z) = \frac{e^z - e^{-z}}{2},$$

$$\tanh(z) = \frac{e^z - e^{-z}}{e^z + e^{-z}}.$$

The three additional functions $\mathrm{sech}(z)$, $\mathrm{csch}(z)$, and $\coth(z)$ are less important; they are the reciprocals of the others as in the case of the trigonometric functions.

▶ **Exercise 2.37.** Find formulas for cosh and sinh in terms of cos and sin.

7. The geometry of addition and multiplication

The algebraic methods used thus far can be made much more appealing by reasoning geometrically. We therefore return to the definitions of addition and multiplication in $\mathbf{C}$ and interpret them geometrically. We have already observed that addition is easy to understand. Given $z, w \in \mathbf{C}$, we think of each as a vector based at 0. Then $z + w$ is the usual vector sum, obtained by the parallelogram law as in Figure 1.5. What does zw represent?

To answer this natural question, we will introduce the polar representation of complex numbers. For $z \neq 0$, we will write $z = |z|e^{i\theta}$ and call this expression the *polar representation* of z. Here $|z|$ is the magnitude of z, namely the length of the vector from 0 to z, and θ is the angle this vector makes with the positive x-axis. If we also write $w = |w|e^{i\phi}$, then we obtain

$$zw = |z|e^{i\theta}|w|e^{i\phi} = |z| \, |w|e^{i(\theta+\phi)}.$$

Thus multiplying two complex numbers multiplies their magnitudes and adds their angles. Two things are happening when we multiply by $|w|e^{i\phi}$. We are dilating, or changing the scale, by a factor of $|w|$, and we are rotating through an angle of ϕ.

We have already defined the complex exponential; the discussion there included some subtle points. In particular our definition of $e^{i\theta}$ involved power series; the reader who feels he or she knows what sine and cosine of an angle are might ask whether the concept of *angle* is precise. Giving the definition of an angle is difficult. For example, θ and $\theta + 2\pi$ are not the same number, but they represent the same angle. Furthermore, most careful definitions of angle rely on being able to compute the length of a circular arc, but defining the length of a circular arc relies on a limiting process.

We next indicate how to find roots of complex numbers by using the polar representation. Even for square roots this method is far simpler than the method from Proposition 2.1. On the other hand, the answers obtained are expressed in the polar form rather than in the form $x + iy$.

Given $w \neq 0$, we wish to find all its k-th roots. To do so, we write $w = |w|e^{i(\theta+2n\pi)}$ for $n = 0, 1, k - 1$. The k-th roots have the following values:

$$z_k = |w|^{\frac{1}{k}} e^{\frac{i}{k}(\theta+2n\pi)}.$$

Since $n < k$, the k values of the angle all lie in an interval of length less than 2π. Hence for $n = 0, 1, ..., k - 1$ the z_k are distinct. When $n = k$, we start repeating the values of the angle. We illustrate by finding the cube roots of $8i$. Start with

$$8i = 8e^{i(\frac{\pi}{2}+2n\pi)}.$$

The three cube roots are $2e^{i\mu}$, where μ can be $\frac{\pi}{6}$, $\frac{5\pi}{6}$, and $\frac{9\pi}{6}$. Expressing these in the form $x + iy$ gives the three solutions $\sqrt{3} + i$, $-\sqrt{3} + i$, and $-2i$.

Most of the subsequent exercises illustrate the power of geometric reasoning.

▶ **Exercise 2.38.** Find all complex z, written $z = x + iy$, such that $z^3 = 1$. Do the same for $z^8 = 1$.

▶ **Exercise 2.39.** Show using the polar form that each nonzero complex number has two distinct square roots.

▶ **Exercise 2.40.** Express the complex cube roots of -1 in the form $x + iy$.

▶ **Exercise 2.41.** Express $-\sqrt{3} + i$ in the polar form $|z|e^{i\theta}$.

▶ **Exercise 2.42.** Take a regular n-gon centered at the origin. Consider the n different vectors from the origin to the vertices. Show that they sum to zero.

▶ **Exercise 2.43.** Assume $\omega = a + ib$ and that $\omega^m = 1$. Express ω^{m-1} in the form $A + iB$. (No computation required!)

▶ **Exercise 2.44.** Let Q be the first quadrant in $\mathbf{C}$. Sketch $\{z \in Q : \text{Im}(z^3) > 0\}$.

▶ **Exercise 2.45.** For each positive integer m, describe geometrically the mapping $z \to z^m$.

▶ **Exercise 2.46.** Suppose $\omega^p = 1$ and that no smaller power of ω equals 1. Find a simple formula for the following expression:

$$1 - \prod_{j=0}^{p-1}(1 - t\omega^j).$$

Suggestion: If you have no idea, guess the answer by trying a few small values of p.

8. Logarithms

The usefulness of complex exponentials and the polar form $z = |z|e^{i\theta}$ suggests that we attempt to introduce complex logarithms. Doing so leads to exciting subtleties and the subject of topology.

Let us first recall some basic facts about the (real) natural logarithm function, defined for positive real numbers t. Perhaps its most appealing definition is the integral:

$$(30) \qquad \log(t) = \int_1^t \frac{du}{u}.$$

From (30) it is evident that $\log(1) = 0$ and that log is an increasing function. It follows for $s \in \mathbf{R}$ that $\log(t) = s$ if and only if $t = e^s$, where e^s is defined by its power series; see Exercise 2.52. The functional equation

$$(31) \qquad \log(ab) = \log(a) + \log(b)$$

follows from either the functional equation for the exponential function or by changing variables in the integral (30) defining the logarithm. We prove (31) by a method we will employ again in Chapter 8 when we study winding numbers. Put $g(x) = \log(ax) - \log(a) - \log(x)$. The functional equation (31) is equivalent to $g(x) = 0$ for all x. By inspection, $g(1) = 0$. To prove (31), it therefore suffices to show that g is constant. To do so, we check that g' vanishes. A simple computation using the fundamental theorem of calculus shows, for all x, that

$$g'(x) = \frac{a}{ax} - \frac{1}{x} = 0.$$

Presuming that (31) holds in general, we try to define the logarithm of a nonzero complex number as follows. Given $z \neq 0$, we write $z = |z|e^{i\theta}$ and put

$$(32) \qquad \log(z) = \log|z| + i\theta.$$

One problem with (32) is that θ could be any value of the angle. All such values differ by multiples of 2π. Hence sometimes we regard the logarithm as a multiple-valued function. In other words, $\log(z)$ means the list of all such values.

▶ **Exercise 2.47.** Using (32), find all values of $\log(e)$, $\log(i)$, and $\log(-1)$.

On the other hand, we would like the logarithm to be a well-defined (single-valued) function. We therefore need some way to pick the angle θ. To do so, we choose a *branch cut*. That is, we draw a half-line connecting 0 and infinity and decree that the logarithm is not defined on this half-line. See Figure 2.4 for two typical branch cuts. Suppose that this half-line makes an angle μ with the positive x-axis, where $0 \le \mu < 2\pi$. We then choose an integer n and decree that the angle θ satisfies $\mu + 2n\pi < \theta < \mu + 2(n+1)\pi$. For each such choice of μ and n we obtain a different function; each such function is called a *branch* of the logarithm. Think of a spiral staircase to help visualize this situation.

▶ **Exercise 2.48.** Find all complex numbers z satisfying $e^{2z} = -1$.

▶ **Exercise 2.49.** Recall that the hyperbolic cosine is defined by the formula $\cosh(z) = \frac{e^z + e^{-z}}{2}$. Find all values of z for which $\cosh(z) = 1$ and all values of z for which $\cosh(z) = 0$. (Suggestion: Work first with the equation $\cosh(z) = \frac{e^z + e^{-z}}{2} = w$, use the quadratic formula, and then put $w = 0$ and $w = 1$.)

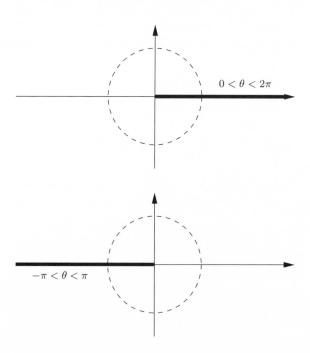

Figure 2.4. Branch cuts for the logarithm.

Given complex numbers z and w, the definition of z^w also requires great care. Again, one approach is to define z^w to be multiple-valued, via the rule

$$z^w = e^{w \log z}.$$

For n a positive integer, finding n-th roots of a complex number by using the polar form works well. We take the usual n-th root of the modulus and obtain for $0 \le k < n$, the n values

(33) $$z^{\frac{1}{n}} = |z|^{\frac{1}{n}} e^{\frac{i\theta + 2\pi k i}{n}}.$$

▶ **Exercise 2.50.** Show, for $z \ne 0$, that the n-th roots in (33) are distinct.

▶ **Exercise 2.51.** Use the polar form to find all fourth roots of -1.

▶ **Exercise 2.52.** Prove that the real logarithm as defined by (30) is the inverse of the real exponential function.

Strange things can happen when we take logarithms or complex powers of complex numbers.

▶ **Exercise 2.53.** What is wrong with the following reasoning suggesting that $2\pi i = 0$?

$$0 = \log(1) = \log((-i)i) = \log(-i) + \log(i) = \frac{3\pi i}{2} + \frac{\pi i}{2} = 2\pi i.$$

▶ **Exercise 2.54.** Define z^w by the multiple-valued formula $z^w = e^{w \log z}$. Find all possible values of i^i and of $(1 - i)^{4i}$.

▶ **Exercise 2.55.** Assume z^w is defined by the multiple-valued formula $z^w = e^{w \log z}$. Show in general that

$$\sqrt{\zeta w} \ne \sqrt{\zeta}\sqrt{w}.$$

▶ **Exercise 2.56.** Find formulas for $\sin^{-1}$, $\cos^{-1}$, and $\tan^{-1}$ using logarithms. Be careful about the branches. Use the quadratic formula as in Exercise 2.49.

▶ **Exercise 2.57.** This extended exercise outlines an alternative approach to complex exponentials and the trigonometric functions. In this approach we avoid power series but we use complex differentiation as in Definition 3.2. See Chapter 5 for additional discussion of the complex derivative. We define the exponential function $f(z)$ to be the unique solution to the differential equation $f'(z) = f(z)$ with $f(0) = 1$. Note that f is then infinitely differentiable. Consider $g(z) = f(iz)$ and $h(z) = f(-iz)$. Use the chain rule to show that $g''(z) = -g(z)$ and $h''(z) = -h(z)$. Put $c(z) = \frac{g(z) + h(z)}{2}$ and put $s(z) = \frac{g(z) - h(z)}{2i}$. Show that s is the unique solution to $s'' = -s$ with $s(0) = 0$ and $s'(0) = 1$. Show that $s' = c$ and that $c' = -s$. Verify by taking derivatives that $c^2 + s^2 = 1$. If z is a real number, show that $c(z)$ and $s(z)$ are real numbers.

▶ **Exercise 2.58.** Use complex numbers to show (for a, b not both 0) that

$$\int e^{ax} \cos(bx) dx = \frac{e^{ax}}{a^2 + b^2} \left(a\cos(bx) + b\sin(bx) \right).$$

Do the same integral using integration by parts (twice).

▶ **Exercise 2.59.** Use complex numbers to show that

$$\int \sec(x) = \log\left(|\sec(x) + \tan(x)| \right).$$

Chapter 3

Complex Numbers and Geometry

In this chapter we study familiar geometric objects in the plane, such as lines, circles, and conic sections. We develop our intuition and results via complex numbers rather than via pairs of real numbers. By the end of the chapter we will follow Riemann and think of complex numbers as points on a sphere where the north pole is the point at infinity.

1. Lines, circles, and balls

We often describe geometric objects via equations. The algebra helps the geometry and the geometry guides the algebra. Two kinds of equations, *parametric equations* and *defining equations*, provide different perspectives on the geometry. Given a complex-valued function f, the set of points p for which $f(p) = 0$ is called the *zero-set* of f. A defining function for a set S is thus a function whose zero-set is precisely S. A parametric equation for a set S is a function whose image is the set S. We will use parametric equations when we compute complex line integrals in Chapter 6. Both kinds of equations arise frequently.

We first consider the Euclidean plane as $\mathbf{R}^2$, but we quickly change perspective and think of the plane as $\mathbf{C}$. Let L be a line in $\mathbf{R}^2$. We can regard L as the set of points (x, y) satisfying an equation of the form

$$(1) \qquad\qquad Ax + By + C = 0,$$

where not both A and B are zero. Equation (1) is called a *defining equation* for L. A point (x, y) lies on L if and only if (x, y) satisfies (1). A defining equation is not unique; we could multiply (1) by a nonzero constant, or even a nonzero function, and the set of solutions would not change. It is natural to seek the simplest possible defining equation; for lines the defining equation should be linear. When $B \neq 0$ in (1), we often solve the defining equation for y and say that the equation of the line is $y = mx + b$; here $m = \frac{-A}{B}$ is the slope of the line. When $B = 0$, we obtain the special case of the line with infinite slope given by x equals a constant.

Alternatively we can describe L via *parametric equations*. Especially in higher dimensions, the parametric approach has many advantages. A line in the plane through the point (x_0, y_0) is determined by its direction vector (u, v); thus L is the set of points $(x_0, y_0) + t(u, v)$ for $t \in \mathbf{R}$. Here the real number t is called a parameter; it is often useful to regard t as *time* and to think of a particle moving along the line. This formulation works in higher dimensions; the parametric equation $\gamma(t) = \mathbf{p} + t\mathbf{v}$ defines a line containing $\mathbf{p}$ and with direction vector $\mathbf{v}$.

We next consider the same issues for circles. A circle with center at (x_0, y_0) and radius R has the defining equation

$$(2) \qquad (x - x_0)^2 + (y - y_0)^2 - R^2 = 0.$$

We could also write the circle using parametric equations:

$$(3) \qquad (x(t), y(t)) = (x_0 + R\cos(t), y_0 + R\sin(t)).$$

Now the parameter t lives (for example) in the interval $[0, 2\pi)$; if we let t vary over a larger set, then we cover points on the circle more than once. Another possible parametrization of a circle will be derived in Chapter 8. There we show that the unit circle can be described by the parametric equations

$$(4) \qquad (x(t), y(t)) = \left(\frac{1 - t^2}{1 + t^2}, \frac{2t}{1 + t^2} \right),$$

where now $-\infty < t < \infty$. We get all points on the unit circle except for $(-1, 0)$, which we realize by allowing t to take the value infinity. Figure 3.1 indicates the geometric meaning of the parameter t.

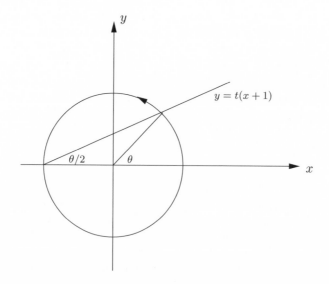

Figure 3.1. Parametrizing the unit circle.

▶ **Exercise 3.1.** Show that (4) parametrizes the unit circle, except for $(-1, 0)$. Show in (4) that $y = t(x + 1)$. What is the geometric meaning of t?

▶ **Exercise 3.2.** If $(x(t), y(t)) = (\cos(\theta), \sin(\theta))$, express the parameter t from (4) in terms of θ. Hint: Look at Figure 3.1. See Chapter 8 for more information.

Things in the plane invariably simplify by using complex variables. A parametric equation for a line in $\mathbf{C}$ is given by $z(t) = z_0 + tv$, where z_0 is a point on the line and v is any nonzero complex number. Note that t is real. The same line has defining equation

$$(5) \qquad\qquad \mathrm{Re}\,((z - z_0)i\overline{v}) = 0.$$

We can derive (5) with almost no computation. First of all, by inspection, z_0 lies on the solution set to (5). Let us reason geometrically. We know that

$$(6) \qquad\qquad |\zeta + w|^2 = |\zeta|^2 + |w|^2 + 2\mathrm{Re}(\zeta\overline{w}).$$

Hence, if $\mathrm{Re}(\zeta\overline{w}) = 0$, then (6) says (the converse of the Pythagorean theorem!) that ζ and w are perpendicular. Therefore (5) states that $(z - z_0)$ is perpendicular to the vector $-iv$. Since multiplication by $\pm i$ is a rotation of $\pm 90°$, we conclude that the vectors $z - z_0$ and v point in the same direction. Thus (5) says both that z_0 is on the line and that v is the direction of the line.

We say a few words about circles and balls. A circle of radius r with center at p is the set of z satisfying $|z - p| = r$. We could also write the circle parametrically as the set of z satisfying $z = p + re^{i\theta}$ for $0 \le \theta < 2\pi$. The closed ball of radius r about p is given by the set of z satisfying $|z - p| \le r$ and the open ball $B_r(p)$ is given by the set of z satisfying $|z - p| < r$. Sometimes, we say *disk* instead of ball. The term *unit disk* means $\{z : |z| < 1\}$. Open balls are important because they lead to the more general notion of *open set*.

Definition 1.1. A subset Ω of $\mathbf{C}$ is called *open* if, for each $z \in \Omega$, there is an $r > 0$ such that $B_r(z) \subset \Omega$.

▶ **Exercise 3.3.** Show that the complement of a closed ball is an open set.

▶ **Exercise 3.4.** Show that the collection $\mathcal{F}$ of open subsets of $\mathbf{C}$ satisfies the following properties:

1) The empty set $\varnothing$ and the whole space $\mathbf{C}$ are elements of $\mathcal{F}$.

2) If A, B are elements of $\mathcal{F}$, then so is $A \cap B$.

3) If A_α is any collection of elements of $\mathcal{F}$, then $\bigcup A_\alpha$ is also in $\mathcal{F}$.

We pause to introduce the definition of a topology. Let X be a set, and let $\mathcal{F}$ be a collection of subsets of X. Then $\mathcal{F}$ is called a *topology* on X, and the pair $(X, \mathcal{F})$ is called a *topological space*, if $\mathcal{F}$ satisfies the three properties from the previous exercise. The elements of $\mathcal{F}$ are called *open subsets*. When (X, δ) is a metric space (Section 6 of Chapter 1), we have already given the definition of open set. The collection of open sets in a metric space does satisfy the three properties that make $(X, \mathcal{F})$ into a topological space.

There are many topologies on a typical set X. For example, we could decree that every subset of X is open. At the other extreme we could decree that the only open subsets of X are the empty set and X itself. The concept of topological space allows one to give the definition and properties of continuous functions solely in terms of the open sets.

We close this section by showing that two specific open subsets of $\mathbf{C}$ can be considered the same from the point of view of complex analysis. The sense in which they are the same is that there is a bijective complex analytic mapping between them. In the next lemma these sets are the open unit ball and the open upper half-plane, defined as the set of z for which $\mathrm{Im}(z) > 0$. See Section 4 of Chapter 8 for these considerations in more generality. While we do not yet wish to develop these ideas, the next lemma also anticipates our study of linear fractional transformations and provides a simple example of conformal mapping.

Lemma 1.1. *Put* $\zeta = i\frac{1-z}{1+z}$. *Then* $|z| < 1$ *if and only if* $\mathrm{Im}(\zeta) > 0$.

Proof. We write $\mathrm{Im}(\zeta)$ as $\frac{\zeta - \overline{\zeta}}{2i}$ and compute:

$$\mathrm{Im}(\zeta) = \frac{\zeta - \overline{\zeta}}{2i} = \frac{1}{2i}\left(i\frac{1-z}{1+z} + i\frac{1-\overline{z}}{1+\overline{z}}\right) = \frac{1}{2}\left(\frac{1-z}{1+z} + \frac{1-\overline{z}}{1+\overline{z}}\right).$$

After clearing denominators, we find that $\mathrm{Im}(\zeta) > 0$ if and only if

$$0 < \frac{1}{2}\big((1-z)(1+\overline{z}) + (1-\overline{z})(1+z)\big) = \frac{1}{2}(2 - 2z\overline{z}) = 1 - |z|^2,$$

that is, if and only if $|z| < 1$. $\qquad\square$

The mapping $z \to i\frac{1-z}{1+z}$ is called a linear fractional transformation. Such transformations map the collection of lines and circles in $\mathbf{C}$ to itself. See Section 4.

2. Analytic geometry

We begin by recalling geometric definitions of hyperbolas, ellipses, and parabolas. We also express these objects using defining equations.

Definition 2.1. A *hyperbola* is the set of points in a plane defined by the following condition. Given distinct foci p and q, the hyperbola consists of those points for which the absolute difference in distances to these two foci is a real nonzero constant. A defining equation is

(7) $$|z - p| - |z - q| = \pm c.$$

An *ellipse* is the set of those points for which the sum of the distances to these foci is a positive constant. A defining equation is

(8) $$|z - p| + |z - q| = c.$$

A circle is the set of points in the plane whose distance to a given point is constant. That distance is called the *radius* of the circle. We include the case of a single point as a circle whose radius is 0. A circle of positive radius is the special case of an ellipse when the foci p and q are equal. A defining equation is then $|z - p| = r$.

A parabola is the set of points in a plane that are equidistant from a given point (the focus) and a given line (the directrix). If the focus is p and the line is given by $z_0 + tv$, then we may take

(9) $$\left(\mathrm{Im}((z - z_0)\overline{v})\right)^2 = |v|^2 |z - p|^2$$

for a defining equation. See Proposition 3.2. We can simplify the equation (9) slightly, because without loss of generality we may assume that $|v| = 1$.

A variant of the definition of a parabola can also be used to define ellipses and hyperbolas. Given a focus and directrix, one considers the set of points for which the distance to the focus is a constant positive multiple $\mathcal{E}$ of the distance to the directrix. This number $\mathcal{E}$ is called the **eccentricity**. When $\mathcal{E} = 1$, the set is a parabola. When $\mathcal{E} < 1$, the set is an ellipse, and when $\mathcal{E} > 1$, the set is a hyperbola. Most calculus books have lengthy discussions of this matter. See for example [**24**]. We will use complex variables to develop a somewhat different intuition.

We mention also that the defining property (8) of an ellipse helps explain why *whispering galleries* work. In such a gallery, one stands at one focus f_1 and whispers into a wall shaped like an ellipsoid. The sound wave emanating from f_1 reflects off the wall and passes through the other focus f_2. Hence someone located at f_2 can hear the whisper from f_1. This property follows from the law of reflection; imagine placing a mirror tangent to the ellipse at the point p on the ellipse hit by the sound wave. The line segment from f_1 to p and the line segment from p to the reflection of f_2 are part of the same line.

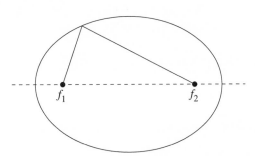

Figure 3.2. The geometric definition of an ellipse.

The definition of a circle allows for a *degenerate* situation consisting of a single point. Degenerate situations for hyperbolas can be more subtle. We pause now to consider such a degenerate situation; Section 3 provides additional examples. Let $H_\pm$ be the set of z such that $|z + 1| - |z - 1| = \pm 2$. Computation (Exercise 3.12) shows that either equation forces z to be real. We can then check that H_+ consists of those real numbers at least 1, and H_- consists of those real numbers at most -1. We can regard these two rays as branches of a degenerate hyperbola.

3. Quadratic polynomials

In this section we develop the relationship between geometry and algebra. For the most part we limit our discussion to quadratic polynomials and the geometry of their zero-sets. We begin however with a few words about polynomials of arbitrary degree in two real variables.

Let $p(x, y)$ be a polynomial with real coefficients in the two real variables x, y. Thus, there are real coefficients c_{ab} such that

$$(10) \qquad p(x, y) = \sum_{a=0}^{m} \sum_{b=0}^{n} c_{ab} x^a y^b.$$

We say that p has degree k if $c_{ab} = 0$ whenever $a + b > k$ and, for some a, b with $a + b = k$, we have $c_{ab} \neq 0$. Let us substitute $x = \frac{z+\overline{z}}{2}$ and $y = \frac{z-\overline{z}}{2i}$ into (10). We obtain a polynomial $\Phi(z, \overline{z})$, defined by

$$(11) \qquad \Phi(z, \overline{z}) = \sum_{a=0}^{m} \sum_{b=0}^{n} c_{ab} \left(\frac{z+\overline{z}}{2}\right)^a \left(\frac{z-\overline{z}}{2i}\right)^b = \sum d_{ab} z^a \overline{z}^b.$$

The coefficients d_{ab} are not in general real, but the values of $\Phi(z, \overline{z})$ are real. Equating coefficients in $\Phi = \overline{\Phi}$ shows, for all indices a, b, that the Hermitian symmetry condition $d_{ab} = \overline{d_{ba}}$ holds. Conversely suppose we are given a polynomial of the form $\Phi(z, \overline{z}) = \sum d_{ab} z^a \overline{z}^b$ and the Hermitian symmetry condition is satisfied. Then $\Phi(z, \overline{z})$ is real for all z. The following definition and proposition clarify the issues.

Definition 3.1. Let $\Phi(z, \overline{w})$ be a polynomial in the two complex variables z and $\overline{w}$. We say that Φ is Hermitian symmetric if for all z and w we have

$$\Phi(z, \overline{w}) = \overline{\Phi(w, \overline{z})}.$$

Proposition 3.1. *Let Φ be the polynomial in two complex variables defined by*

$$\Phi(z, \overline{w}) = \sum d_{ab} z^a \overline{w}^b.$$

The following statements are equivalent:

- *Φ is Hermitian symmetric.*
- *For all a, b we have $d_{ab} = \overline{d_{ba}}$.*
- *For all z, $\Phi(z, \overline{z})$ is real.*

Proof. This simple proof is left to the reader.   □

To each real polynomial in x, y there corresponds a unique Hermitian symmetric polynomial in $z, \overline{w}$, and conversely each Hermitian symmetric polynomial in $z, \overline{w}$ defines a real polynomial after setting $\overline{w}$ equal to $\overline{z}$. It therefore makes little difference whether we study real polynomials or Hermitian symmetric polynomials. The Hermitian symmetric perspective seems easier to understand.

▶ **Exercise 3.5.** Prove Proposition 3.1.

▶ **Exercise 3.6.** Prove that the correspondence going from (10) to (11) between real and Hermitian symmetric polynomials preserves degree.

Our discussion of analytic geometry leads to quadratic polynomials, namely those of (total) degree two. We must distinguish the total degree from the degree in z alone. For example, the Hermitian symmetric polynomial $|z|^2 = z\overline{z}$ is quadratic; its degree is two, but it is of degree one in the z variable alone.

For hyperbolas and ellipses we start with the defining relations $|z-p| = c \pm |z-q|$ and square both sides. Then we isolate the term involving $|z-q|$ and square again. After simplifying, only terms of degree at most two remain. For a parabola, we must use the formula (9), which follows from the following result:

Proposition 3.2. *Let L be the line in $\mathbf{C}$ defined by $z(t) = z_0 + tv$. The (minimum) distance δ from a point z to this line L is given by*

$$(12) \qquad \delta = |\frac{1}{v}\text{Im}((z-z_0)\overline{v})| = |\frac{(z-z_0)\overline{v} - \overline{(z-z_0)v}}{2\overline{v}}|.$$

Proof. Consider the squared distance $f(t)$ from z to a point on L. Thus $f(t) = |z - z_0 - tv|^2$. We expand and find the minimum of the quadratic polynomial f by calculus. We obtain the equations

$$f(t) = |z-z_0|^2 - 2t\text{Re}((z-z_0)\overline{v}) + t^2|v|^2,$$
$$f'(t) = -2\text{Re}((z-z_0)\overline{v}) + 2t|v|^2.$$

Therefore the minimum occurs when

$$t = \frac{\text{Re}((z-z_0)\overline{v})}{|v|^2}.$$

Plugging in this value for t in f gives the minimum squared distance, namely

$$\delta^2 = \left|(z-z_0) - \frac{\text{Re}((z-z_0)\overline{v})}{\overline{v}}\right|^2 = \left|\frac{2(z-z_0)\overline{v} - \overline{(z-z_0)}v - (z-z_0)\overline{v}}{2\overline{v}}\right|^2$$
$$= \left|\frac{1}{\overline{v}}\text{Im}((z-z_0)\overline{v})\right|^2.$$

$\square$

▶ **Exercise 3.7.** Let L be the line in $\mathbf{C}$ defined by $z(t) = z_0 + tv$. Verify (9) by using Proposition 3.2.

▶ **Exercise 3.8.** Find, in terms of x and y, the equation of a parabola with focus at $(3, 0)$ and directrix the line $x = 1$.

▶ **Exercise 3.9.** Find, in terms of x and y, the equation of any hyperbola with foci at $\pm 3i$.

▶ **Exercise 3.10.** What object is defined by the condition that the eccentricity is 0? What object is defined by the condition that the eccentricity is infinite?

Viewing these objects via eccentricity enables us to conceive of hyperbolas, parabolas, and ellipses in similar fashions. These objects, as well as points, lines, and pairs of lines, are zero-sets for Hermitian symmetric quadratic polynomials.

Proposition 3.3. *Let $z_0 + tv$ define a line in $\mathbf{C}$ and let $p \in \mathbf{C}$. Define a family of Hermitian symmetric polynomials $\Psi_{\mathcal{E}}$ by the formula*

$$\Psi_{\mathcal{E}} = |z-p|^2 - \mathcal{E}^2|\frac{(z-z_0)\overline{v} - \overline{(z-z_0)v}}{2\overline{v}}|^2.$$

Then the zero-set of $\Psi_{\mathcal{E}}$ is an ellipse for $0 < \mathcal{E} < 1$, a parabola for $\mathcal{E} = 1$, and a hyperbola for $\mathcal{E} > 1$.

Proof. The conclusion follows from (12) and the characterizations of the objects using eccentricities. Alternatively, one can derive the statement from the subsequent discussion by computing the determinant Δ. $\square$

Consider the most general Hermitian symmetric quadratic polynomial:

$$(13) \qquad \Phi(z, \overline{z}) = \alpha z^2 + \overline{\alpha z}^2 + \beta z + \overline{\beta}\,\overline{z} + \gamma z\overline{z} + F = 0.$$

In (13), α and β are complex, whereas γ and F are real. We will analyze the zero-sets of such polynomials, thereby providing a complex variables approach to conic sections. The analysis requires many cases; we first assume $\alpha = 0$ in (13). The zero-set $\mathbf{V}$ of Φ must then be one of the following objects: the empty set, a line, a point, a circle, all of $\mathbf{C}$. This situation arises again in Theorem 4.1 when we study linear fractional transformations.

- Assume $\alpha = \beta = \gamma = 0$ in (13). The equation $\Phi = 0$ becomes $F = 0$. Thus $\mathbf{V}$ is empty if $F \neq 0$ and $\mathbf{V}$ is all of $\mathbf{C}$ if $F = 0$.

- Assume $\alpha = \gamma = 0$ in (13) but $\beta \neq 0$. The equation $\Phi = 0$ becomes

$$2\mathrm{Re}(z\beta) + F = 0.$$

Hence in this case $\mathbf{V}$ is a line.

- Assume $\alpha = 0$ and $\gamma \neq 0$. We proceed analogously to the proof of the quadratic formula. We complete the square in the equation $\Phi = 0$ to get

$$(14) \qquad |z + \frac{\overline{\beta}}{\gamma}|^2 = \frac{|\beta|^2 - F\gamma}{\gamma^2}.$$

Hence $\mathbf{V}$ is either empty, a point, or a circle; the answer depends on whether the right-hand side of (14) is negative, zero, or positive.

It remains to discuss the case where $\alpha \neq 0$, in which case the defining equation must be quadratic. The expression $\Delta = \gamma^2 - 4|\alpha|^2$ governs the geometry. When $\Delta < 0$, we get a (possibly degenerate) hyperbola; when $\Delta = 0$, we get a (possibly degenerate) parabola; when $\Delta > 0$, we get a (possibly degenerate) ellipse. Below we will interpret Δ from the point of view of Hermitian symmetric polynomials.

We first give some examples of degenerate situations arising when $\alpha \neq 0$.

- The equation $\mathrm{Re}(z^2) = 0$, where $\Delta = -1$, defines two lines rather than a hyperbola.

- The equation $(z + \overline{z})^2 = 0$, where $\Delta = 0$, defines a line rather than a parabola.

- The equation $z^2 + \overline{z}^2 + 2|z|^2 - 2(z + \overline{z}) = 0$, where $\Delta = 0$, defines two lines rather than a parabola.

- The equation $|z|^2 = 0$, where $\Delta = 1$, defines a point rather than an ellipse.

Notice that two lines can be a degenerate version of either a parabola or a hyperbola. Another interesting point is that sometimes one line should be regarded as one line, but other times it should be regarded as two lines! The linear equation $x = 0$ defines the line $x = 0$ once, and the zero-set should be regarded as one line. On the other hand, the quadratic equation $x^2 = 0$ defines the single line $x = 0$ twice, and the zero-set should be regarded as two lines.

To determine what kind of an object (13) defines can be a nuisance because of degenerate cases. In the very degenerate case where no quadratic terms are present ($\alpha = \gamma = 0$) the linear terms determine whether the object is a line, the empty set, or all of **C**. When the quadratic part is not identically zero, the linear terms usually amount to translations and do not matter; the exception comes when the quadratic part itself is degenerate and the linear terms determine whether the object is a parabola. For example compare the equations $x^2 - 1 = 0$ and $x^2 - y = 0$. The first defines two lines while the second defines a parabola.

We consider the pure quadratic case from the complex variable point of view. Thus we let

$$(15) \qquad \Phi(z, \overline{z}) = \alpha z^2 + \overline{\alpha}\overline{z}^2 + \gamma z\overline{z}.$$

For a positive constant c the set $\Phi = c$ defines an ellipse whenever $\Phi(z, \overline{z}) > 0$ for $z \neq 0$. After dividing by $|z|^2$ and introducing polar coordinates, we find that the condition becomes

$$(16) \qquad \alpha e^{2i\theta} + \overline{\alpha} e^{-2i\theta} + \gamma > 0.$$

The minimum value of the left-hand side of (16) occurs when $\alpha e^{2i\theta} = -|\alpha|$. Hence the condition for being an ellipse is that $\gamma - 2|\alpha| > 0$.

One can obtain this inequality by other methods. We can write (15) in terms of x, y as

$$P(x, y) = (\gamma + 2\mathrm{Re}(\alpha))x^2 - 4\mathrm{Im}(\alpha)xy + (\gamma - 2\mathrm{Re}(\alpha))y^2,$$

or in terms of matrices as

$$M = \begin{pmatrix} \gamma + 2\mathrm{Re}(\alpha) & -2\mathrm{Im}(\alpha) \\ -2\mathrm{Im}(\alpha) & \gamma - 2\mathrm{Re}(\alpha) \end{pmatrix}.$$

The polynomial P and the matrix M are equivalent ways of defining a quadratic form. The behavior of this quadratic form is governed by the eigenvalues of M. Their product is the determinant Δ, given in (17), and their sum is the trace 2γ.

$$(17) \qquad \Delta = \gamma^2 - 4(\mathrm{Re}(\alpha))^2 - 4(\mathrm{Re}(\beta))^2 = \gamma^2 - 4|\alpha|^2.$$

Things degenerate when $\Delta = 0$. When $\Delta > 0$, we also must have $\gamma > \pm 2\mathrm{Re}(\alpha)$, and both eigenvalues are positive. We then obtain an ellipse for the set $\Phi = c$, for $c > 0$. When $\Delta < 0$, we obtain a hyperbola, where we allow the possibility of two crossing lines.

3.1. The situation using real variables. For comparative purposes we recall how this discussion from elementary analytic geometry proceeds when we stay within the realm of real variables. Consider a polynomial P of degree at most two with real coefficients in the variables x and y. Thus there are real numbers A, B, C, D, E, F such that

$$(18) \qquad P(x, y) = Ax^2 + 2Bxy + Cy^2 + Dx + Ey + F.$$

The set of points (x, y) for which $P(x, y) = 0$ (called the zero-set of P) must be one of the following geometric objects: the empty set, all of $\mathbf{R}^2$, a point, a line, two lines, a circle, a parabola, a hyperbola, or an ellipse. We may regard a circle as a special case of an ellipse. The reader should be able to solve the following exercise.

▶ **Exercise 3.11.** For each of the geometric objects in the above paragraph, give values of the constants A, B, C, D, E, F such that the zero-set of P is that object. Show that all the above objects, except for the entire plane, are zero-sets of polynomials of degree two. Thus even the *degenerate* cases of empty set, point, line, and two lines are possible for the zero-sets of quadratic polynomials in two real variables.

It is possible to completely analyze the possibilities for the zero-set of P; with the proper background this analysis is concise. Without that background things seem messy. We recall the answer.

Again the difficulty in the analysis arises from the many degenerate cases. If $P(x, y) = F$ is a constant, then the zero-set is either everything or nothing (the empty set), according to whether $F = 0$ or not. If P is of degree one, that is, $P(x, y) = Dx + Ey + F$ and at least one of D and E is not zero, then the zero-set is a line. If P is of degree two, then things depend on the expression $AC - B^2$. If $AC - B^2 < 0$, then the object is a hyperbola, with two lines a possibility. If $AC - B^2 > 0$ and $A > 0$, then the object is either empty, a single point, a circle, or an ellipse. We may regard a single point or a circle as a kind of ellipse. We next analyze the possibilities when $AC - B^2 = 0$, still assuming that P has degree two. Note then that at least one of A and C is nonzero, and they cannot have opposite signs. After multiplying P by -1, we may assume that A and C are nonnegative. We eliminate B and write

$$Ax^2 + 2\sqrt{AC}xy + Cy^2 + Dx + Ey + F = (\sqrt{A}x + \sqrt{C}y)^2 + Dx + Ey + F.$$

Analyzing the possible zero-sets is amusing. If $D = E = 0$, then we get

$$(\sqrt{A}x + \sqrt{C}y)^2 + F = 0,$$

which gives the empty set if $F > 0$, a single line if $F = 0$, and a pair of lines if $F < 0$. If at least one of A and D is nonzero, then we write $u = \sqrt{A}x + \sqrt{C}y$ and $v = Dx + Ey$. First assume that v is not a constant multiple of u. Then the equation $P = 0$ becomes $u^2 + v + F = 0$, which defines a parabola. If v is a multiple of u, then the resulting object becomes two lines, one line, or the empty set.

3.2. Back to complex variables. To complete our discussion, we recall how to rewrite everything in complex notation. We use the formulas for x and y in terms of z and $\overline{z}$ to obtain

(19) $0 = P(x, y)$

$$= A(\frac{z + \overline{z}}{2})^2 + 2B(\frac{z + \overline{z}}{2})(\frac{z - \overline{z}}{2i}) + C(\frac{z - \overline{z}}{2i})^2 + D(\frac{z + \overline{z}}{2}) + E(\frac{z - \overline{z}}{2i}) + F.$$

Simplifying (19) gives, for complex numbers α, β, real number γ, and the same real number F, our familiar formula (13).

▶ **Exercise 3.12.** What kind of object does each of the following equations define?

- $iz^2 - i\overline{z}^2 = 4$.
- $|z|^2 + z^2 + \overline{z}^2 = 3$.
- $|z - 1| + |z - 3| = 2$. Also, write this equation in the form (13).
- $\alpha z^2 + \overline{\alpha z}^2 + |z|^2 = 1$. The answer depends on α.

- $z^2 + \overline{z}^2 = 0$.
- $|z + 1| - |z - 1| = \pm 2$. (Be careful!)

Complex analysis offers much geometric information. Figure 3.3 shows that the level sets of the real and imaginary parts of z^2 form orthogonal (perpendicular) hyperbolas; more generally we shall see that the level sets of the real and imaginary parts of a complex analytic function form orthogonal trajectories. This fact lies at the foundation of various applications to physics and engineering. We close this section by posing some related exercises.

▶ **Exercise 3.13.** Let $f(z) = z^2$. What are the real and imaginary parts of f in terms of x, y? Graph their level sets; show that one gets orthogonal hyperbolas.

▶ **Exercise 3.14.** Let $f(z) = z^3$. What are the real and imaginary parts of f in terms of x, y? Can you prove that the corresponding level sets are orthogonal?

▶ **Exercise 3.15.** What are the real and imaginary parts of $\frac{1}{z}$? (Assume $z \neq 0$.) Show by computation that their level sets form orthogonal trajectories. Do the same for $\frac{1}{z^n}$.

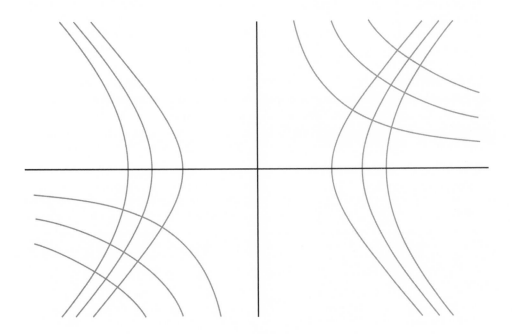

Figure 3.3. Orthogonal hyperbolas.

The close relationship between the trigonometric functions and the unit circle has been useful for us. The next exercise reveals a similar relationship between the hyperbolic functions and the hyperbola $x^2 - y^2 = 1$.

▶ **Exercise 3.16.** Show that

$$\cosh^2(z) - \sinh^2(z) = 1.$$

Find parametric equations for (a branch of) the hyperbola $x^2 - y^2 = 1$.

▶ **Exercise 3.17.** Find formulas for the multi-valued functions $\cosh^{-1}$ and $\sinh^{-1}$ using logarithms. Do the same for $\tanh^{-1}$.

4. Linear fractional transformations

This section glimpses the interesting and important subject of conformal mapping by way of a natural collection of complex analytic functions. We first consider the rational function f given by

$$(20) \qquad\qquad f(z) = \frac{az + b}{cz + d},$$

where a, b, c, d are complex numbers. We cannot allow both c and d to be 0, or else we are dividing by 0. There is another natural condition; we do not want this mapping to degenerate into a constant. Such degeneration occurs if a is a multiple of c and b is the same multiple of d, in other words, if $ad - bc = 0$. By restricting to the case when $ad - bc \neq 0$, we avoid both problems.

Definition 4.1. A *linear fractional transformation* is a rational function of the form (20) where $ad - bc \neq 0$.

We let L denote the collection of linear fractional transformations. We will next show that L can be regarded as the collection of two-by-two complex matrices with determinant one.

Given a linear fractional transformation, we obtain the same function if we multiply the numerator and denominator by the same nonzero constant. We may therefore assume, without loss of generality, that $ad - bc = 1$. Assume f is as in (20) and that $ad - bc = 1$. We can record the information defining f as a two-by-two matrix

$$F = \begin{pmatrix} a & b \\ c & d \end{pmatrix}.$$

Thinking of F rather than of f has some advantages. The identity map corresponds to the identity matrix, and composition corresponds to matrix multiplication. If $f, g \in L$, then so is $g \circ f$. Write $g(z) = \frac{Az + B}{Cz + D}$. We compute the composition

$$(21) \quad g(f(z)) = \frac{Af(z) + B}{Cf(z) + D} = \frac{A\frac{az+b}{cz+d} + B}{C\frac{az+b}{cz+d} + D} = \frac{(Aa + Bc)z + (Ab + Bd)}{(Ca + Dc)z + (Cb + Dd)}.$$

We identity f and g with matrices

$$F = \begin{pmatrix} a & b \\ c & d \end{pmatrix},$$

$$G = \begin{pmatrix} A & B \\ C & D \end{pmatrix}.$$

Then the composition $g \circ f$ is identified with the matrix

$$(22) \qquad GF = \begin{pmatrix} Aa + Bc & Ab + Bd \\ Ca + Dc & Cb + Dd \end{pmatrix} = \begin{pmatrix} A & B \\ C & D \end{pmatrix} \begin{pmatrix} a & b \\ c & d \end{pmatrix}.$$

Note that $ad - bc$ is the determinant of the matrix F corresponding to f and that $AD - BC$ is the determinant of the matrix G corresponding to g. Since the determinant of the product of matrices is the product of the respective determinants, it follows that the determinant of GF is not zero. Had we assumed each determinant was 1, the product would also be 1. We also obtain a formula for the inverse mapping; we take the inverse matrix. See Exercise 3.18. The natural assumption that the determinant equals 1 enables us to compute the inverse easily:

$$(23) \qquad \begin{pmatrix} a & b \\ c & d \end{pmatrix}^{-1} = \begin{pmatrix} d & -b \\ -c & a \end{pmatrix}.$$

▶ **Exercise 3.18.** Put $w = f(z) = \frac{az+b}{cz+d}$, where $ad - bc = 1$. Solve for z as a function of w. Check that the answer agrees with the inverse matrix from (23).

We may therefore consider L to be the group of two-by-two matrices with complex entries and determinant one. In elementary linear algebra we learn Gaussian elimination, or row operations, as a method for solving a system of linear equations. The effect of row operations is to write a matrix of coefficients as a product of particularly simple matrices. We can do the same thing for linear fractional transformations, and we will obtain a beautiful geometric corollary.

First we discuss the geometric interpretation of the three simplest linear fractional transformations. The mapping $z \to z + \beta = T_\beta(z)$ is a translation. For $\alpha \neq 0$, the map $z \to \alpha z = M_\alpha z$ is a dilation and a rotation; it changes the scale by a factor of $|\alpha|$ and rotates through an angle θ if $\alpha = |\alpha|e^{i\theta}$. The mapping $z \to \frac{1}{z} = R(z)$ is an inversion (taking the reciprocal). For the moment we write T for any translation, M for any multiplication, and R for the reciprocal. We will show that every linear fractional transformation can be written as a composition of these three simpler kinds.

For convenience we write the transformations as matrices:

$$T_\beta = \begin{pmatrix} 1 & \beta \\ 0 & 1 \end{pmatrix},$$

$$M_\alpha = \begin{pmatrix} \alpha & 0 \\ 0 & 1 \end{pmatrix},$$

$$R = \begin{pmatrix} 0 & 1 \\ 1 & 0 \end{pmatrix}.$$

Let $f(z) = \frac{az+b}{cz+d}$. If $c = 0$, there are three possibilities for f. Since $ad - bc \neq 0$, necessarily $d \neq 0$ and $a \neq 0$. In this case $f(z) = \frac{a}{d}z + \frac{b}{d}$. If $b = 0$, then f is a multiplication. If $b \neq 0$, then f is an affine transformation. If $\frac{a}{d} = 1$, f is a translation. Otherwise f is the composition of a translation and a multiplication. Thus, when $c = 0$, the possibilities are $f = M$, $f = T$, and $f = TM$. We may regard the identity mapping as either the translation T_0 or the multiplication M_1.

If $c \neq 0$, we will need to take an inversion. To simplify f, we simply divide $cz + d$ into $az + b$, obtaining $\frac{a}{c}$ with a remainder of $\frac{bc-ad}{c}$. Using the definition of division, we obtain

$$(24) \qquad \frac{az+b}{cz+d} = \frac{a}{c} + \frac{bc-ad}{c(cz+d)}.$$

We interpret (24) as a composition of mappings:

$$z \to cz \to cz + d \to \frac{1}{cz+d} \to \frac{bc-ad}{c}\frac{1}{cz+d} \to \frac{bc-ad}{c}\frac{1}{cz+d} + \frac{a}{c} = \frac{az+b}{cz+d}.$$

Using the above geometric language, we have written

(25) $$f = TMRTM,$$

or more specifically

(26) $$\frac{az+b}{cz+d} = T_{\frac{a}{c}} M_{\frac{bc-ad}{c}} R T_d M_c(z).$$

The respective translations in (26) are not needed when $d = 0$ or $a = 0$, but the notation (25) is still valid because T_0 is the identity mapping.

The next theorem summarizes these results and includes a beautiful consequence. Let $\mathcal{S}$ denote the collection of lines and circles in $\mathbf{C}$. We include the special case of a single point (a circle of radius 0) in $\mathcal{S}$, but we will not fully understand this situation until we deal with infinity in the next section.

The statement of Theorem 4.1 means that the image of a line under a linear fractional transformation is either a line or a circle, and the image of a circle under a linear fractional transformation is a line or a circle. The following example, noted earlier in Lemma 1.1, shows that the image of a circle need not be a line, and conversely. The image of a point is a point, but we need to allow the point at infinity.

Example 4.1. Put $f(z) = i\frac{1-z}{1+z}$. Then $|z| = 1$ if and only if $\text{Im}(f(z)) = 0$. Thus the image under f of the unit circle is the real axis. Then f^{-1} maps the real axis to the unit circle.

We can give a simple description of $\mathcal{S}$ in terms of Hermitian symmetric polynomials. Let A, C be real numbers, and let β be a complex number. Consider the Hermitian symmetric polynomial $\Phi(z, \overline{z})$ defined by

(27) $$\Phi(z, \overline{z}) = A|z|^2 + \beta z + \overline{\beta z} + C.$$

Here the coefficients A and C are real and $\beta \in \mathbf{C}$. We assume that not all of A, β, C are zero. If either A or β is not zero, then Φ is nonconstant, and its zero-set of Φ is either a line or a circle, where we allow the special case of a single point. In case $A = \beta = 0$ but $C \neq 0$, we will think of the zero-set of Φ as a single point at infinity. Conversely, each line or circle is the zero-set of some such Φ.

Theorem 4.1. *Each $f \in L$ maps $\mathcal{S}$ to itself.*

Proof. It is evident that each translation T and each multiplication M maps $\mathcal{S}$ to itself. Thus a composition of such does the same. The conclusion therefore holds when $f = TM$. We next check that the inversion (reciprocal) maps $\mathcal{S}$ to itself. Assume that V is a line or a circle. Then V is the zero-set of some Φ as in (27). Note that $0 \in V$ if and only if $C = 0$. Set $z = \frac{1}{w}$ in (27) and clear denominators. We obtain a new Hermitian symmetric polynomial $\Phi^*(w, \overline{w})$ defined by

$$\Phi^*(w, \overline{w}) = |w|^2 \Phi\left(\frac{1}{w}, \frac{1}{\overline{w}}\right) = A + \beta\overline{w} + \overline{\beta}w + C|w|^2.$$

The zero-set of Φ^* is a circle when $C \neq 0$, and it is a line if $C = 0$. The special case when $\beta = C = 0$ must be considered. Then $A \neq 0$, and the zero-set of Φ^* becomes the point at infinity. We summarize as follows. If Φ is Hermitian symmetric and its zero-set is a line or circle of positive radius, then the same is true for Φ^*. If the zero-set of Φ is a single point p not the origin, then the zero-set of Φ^* is the single point $\frac{1}{p}$. If the zero-set of Φ is the origin, then the zero-set of Φ^* is the point at infinity. Conversely if the zero-set of Φ is the point at infinity, then the zero-set of Φ^* is the origin.

These remarks show that inversion maps S to itself. The same is true for translation and multiplication. Hence (25) implies that every linear fractional transformation maps S to itself. $\qquad\qquad\qquad\qquad\qquad\qquad\qquad\qquad\quad\square$

▶ **Exercise 3.19.** Find a linear fractional transformation that maps the exterior of a circle of radius 2 with center at 2 to the interior of the unit circle.

▶ **Exercise 3.20.** Let f be a given linear fractional transformation. Determine which lines are mapped to circles under f and which circles are mapped to lines.

▶ **Exercise 3.21.** Given a line in $\mathbf{C}$, describe precisely which linear fractional transformations map this line to a circle. Given a circle in $\mathbf{C}$, describe precisely which linear fractional transformations map this circle to a line.

▶ **Exercise 3.22.** Find all linear fractional transformations mapping the real line to itself.

▶ **Exercise 3.23.** Find all linear fractional transformations mapping the unit circle to itself.

▶ **Exercise 3.24.** Show that the conjugation map $z \to \overline{z}$ maps S to itself. Conclude that the mapping $z \to \frac{a\overline{z}+b}{c\overline{z}+d}$ maps S to itself.

5. The Riemann sphere

We are ready to discuss infinity. Riemann had the idea of adding a point to $\mathbf{C}$, called the point at infinity, and then visualizing the result as a sphere. The resulting set is called either the *Riemann sphere* or the *extended complex plane*. The Riemann sphere provides some wonderful new perspectives on complex analysis, and we briefly describe some of these now.

We can realize the Riemann sphere in the following way. Let U_0 be a copy of $\mathbf{C}$. Let U_1 consist of the set of reciprocals of elements in $\mathbf{C}$, where we denote the reciprocal of 0 by ∞. Note that the intersection of these two sets is $\mathbf{C}^*$, standard notation for the nonzero complex numbers. We let $X = U_0 \cup U_1$. When we are working near 0, we work in U_0; when we are working near ∞, we work in U_1. If we are working somewhere and we wish to pass between the two sets, we take a reciprocal. This simple idea leads to the notion of a *Riemann surface* and more generally to that of a *complex manifold*. The subject of *complex geometry* is based upon the study of complex manifolds, but it is far more sophisticated than the geometry we study in this book.

The procedure from the previous section, where we replaced a Hermitian symmetric polynomial Φ with Φ^*, amounts to using the map $z \to \frac{1}{z}$ to pass from U_0

to U_1. The same idea enables us to define limits on the Riemann sphere. Infinity behaves the same as any other point! We have already seen the definition of convergent sequence on $\mathbf{C}$. We recall the definition of limits on $\mathbf{C}$ in order to extend the definition to the Riemann sphere.

Definition 5.1. Fix $a, L \in \mathbf{C}$. Let S be an open set containing a. We say that $\lim_{z \to a} f(z) = L$ if, for all $\epsilon > 0$, there is a $\delta > 0$ such that

(28) $0 < |z - a| < \delta$ implies $|f(z) - L| < \epsilon$.

We can extend the definition of a limit to allow both a and L to be ∞. We can also talk about neighborhoods of infinity.

Definition 5.2. Fix $L \in \mathbf{C}$. Then $\lim_{z \to \infty} f(z) = L$ if and only if $\lim_{z \to 0} f(\frac{1}{z}) = L$. Also, $\lim_{z \to a} f(z) = \infty$ if and only if $\lim_{z \to a} \frac{1}{f(z)} = 0$.

For example, if $k \in \mathbf{N}$, then $\lim_{z \to \infty} z^k = \infty$. Now that we have understood limits, we let X denote $\mathbf{C} \cup \{\infty\}$ together with the topology determined by these limits. We call X the Riemann sphere. By *neighborhood* of a point $p \in \mathbf{C}$ we mean any subset which contains an open ball about p. As we did for limits, we define *neighborhood of infinity* by taking reciprocals.

Definition 5.3. Let S be a subset of the Riemann sphere containing ∞. We say that S is a neighborhood of ∞ if the set of reciprocals of elements of S is a neighborhood of 0.

▶ **Exercise 3.25.** Use Definition 5.2 to show that $\lim_{z \to \infty} \frac{az+b}{cz+d} = \frac{a}{c}$.

▶ **Exercise 3.26.** Use Definition 5.2 to show that $\lim_{z \to \frac{-d}{c}} \frac{az+b}{cz+d} = \infty$.

Once we have understood limits on the Riemann sphere X, we can introduce open sets and make X into a topological space. The resulting space is the basic example of a compact Riemann surface.

The next exercise uses *stereographic projection* to provide a one-to-one correspondence between the unit sphere and the extended complex plane.

▶ **Exercise 3.27.** Consider $\mathbf{R}^3$ with coordinates (x_1, x_2, x_3). Let p be a point on the unit sphere in $\mathbf{R}^3$ other than the north pole $(0, 0, 1)$. Find the line from the north pole to p and see where that point intersects the plane defined by $x_3 = 0$. Call the point of intersection $s(p)$. Define s of the north pole to be infinity. This mapping s is called stereographic projection. Write explicit formulas for $s(p)$ and show that s maps the sphere bijectively onto the extended complex plane. Find a formula for s^{-1}.

▶ **Exercise 3.28.** Find the image of the equator under stereographic projection.

▶ **Exercise 3.29.** Using the notation of Exercise 3.27, assume that $s(p) = z$. Find the point q for which $s(q) = \frac{1}{z}$.

▶ **Exercise 3.30.** Replacing x by $\frac{1}{x}$ is sometimes useful on the real line. For real numbers a, b consider the integral $F(a, b)$ given by

$$F(a, b) = \int_0^\infty \frac{1}{x^2 + 1} \frac{x^b - x^a}{(1 + x^a)(1 + x^b)} dx.$$

First show that $F(a, b) = 0$. Then use this result to show that

$$\int_0^\infty \frac{1}{x^2 + 1} \frac{1}{(1 + x^a)} dx = \frac{1}{2} \int_0^\infty \frac{1}{x^2 + 1} dx = \frac{\pi}{4}.$$

▶ **Exercise 3.31.** (Difficult) Suppose that f is a one-to-one continuous mapping from the Riemann sphere onto itself and that f maps $\mathcal{S}$ to itself. Show that f is either a linear fractional transformation or the conjugate of a linear fractional transformation.

Chapter 4

Power Series Expansions

Power series play a major role in many fields of mathematics. The basic objects of interest in complex analysis are complex analytic functions; these functions are precisely those functions that, near each point of their domains, can be defined by convergent power series. While power series might seem to be a difficult topic, we gain insight by making analogies with the decimal expansion of a real number.

Each real number x in the interval $(0, 1)$ has a decimal expansion; we write $x = \sum_{n=1}^{\infty} a_n (\frac{1}{10})^n$, where each a_n is an integer with $0 \le a_n \le 9$. Power series are analogous; we allow the a_n to be arbitrary complex numbers and we replace $\frac{1}{10}$ with a variable z. The resulting series will not necessarily converge. We use the term *formal power series* for objects of the form

$$\sum_{n=0}^{\infty} a_n z^n,$$

where the coefficients a_n are arbitrary complex numbers and hence we make no assumption on convergence. Even formal power series will be useful for us. We can add and multiply formal power series in the expected manner.

In this chapter we discuss the fundamental issues about convergent power series in one complex variable and we give some striking applications. We study the geometric series in detail, both for its own sake and in preparation for the main developments in Chapter 6. We study the Fibonacci numbers, we give a formula for the sum of the first N p-th powers, and we give a test for when a power series represents a rational function.

Later in this book we will show that all complex analytic functions are given locally by convergent power series expansions. In this chapter we have some fun.

1. Geometric series

The geometric series arises with varying levels of importance throughout mathematics and science; it even gets mentioned in elementary economics courses, when one discusses the *multiplier effect*. In complex analysis, however, the geometric series plays a dominant role. The basic objects of interest in complex analysis are complex analytic, or holomorphic, functions. Such functions can be developed in convergent power series expansions; the key theoretical step in the proof is the Cauchy integral formula, which shows how to reduce the general case to that of the geometric series. We therefore begin with the geometric series.

Consider first the finite geometric series:

$$S_N(z) = \sum_{n=0}^{N} z^n.$$

In Chapter 2, we noted for $z \neq 1$ that

$$(1) \qquad\qquad S_N(z) = \frac{1 - z^{N+1}}{1 - z}.$$

By Definition 4.1 of Chapter 2 we must investigate the limit of S_N as N tends to infinity. We saw there that $\lim_{N \to \infty} S_N(z) = \frac{1}{1-z}$ whenever $|z| < 1$. Hence, for $|z| < 1$, we have the geometric series:

$$(2) \qquad\qquad \sum_{n=0}^{\infty} z^n = \frac{1}{1 - z}.$$

It is remarkable how many explicit series expansions can be found (either formally or rigorously) by manipulating the geometric series. We give some examples involving substitution and the techniques of calculus. The examples involving derivatives and integrals can be fully justified using Corollary 5.1 of Chapter 2.

Example 1.1. By substitution, the series for $\frac{1}{1+z^2}$ is given for $|z| < 1$ by

$$(3) \qquad\qquad \frac{1}{1 + z^2} = \frac{1}{1 - (-z^2)} = \sum_{n=0}^{\infty} (-1)^n z^{2n}.$$

Example 1.2. We expand $\frac{1}{1-z}$ in powers of $(z - p)$ as follows. If $\left|\frac{z-p}{1-p}\right| < 1$, then

$$(4) \qquad \frac{1}{1 - z} = \frac{1}{1 - p - (z - p)} = \frac{1}{1 - p} \frac{1}{1 - \frac{z-p}{1-p}} = \frac{1}{1 - p} \sum_{n=0}^{\infty} \left(\frac{z - p}{1 - p}\right)^n.$$

The last step in (4) follows by substitution into the geometric series. Thus, if $p \in \mathbf{C}$ and $p \neq 1$, we have a series expansion centered about p. Whenever $\left|\frac{z-p}{1-p}\right| < 1$,

$$(5) \qquad\qquad \frac{1}{1 - z} = \sum_{n=0}^{\infty} \left(\frac{1}{1 - p}\right)^{n+1} (z - p)^n.$$

We mention that (5) plays a crucial role in the Cauchy theory from Chapter 6. Below we discuss in detail where a given series converges and its *radius of convergence*. Figure 4.1 indicates why the radius of convergence of (5) is $|p - 1|$, namely the distance from p to the singularity at 1.

Example 1.3 (Series for (a branch of) the inverse tangent function). By Corollary 5.1 of Chapter 2 we can integrate a power series term by term within the circle of

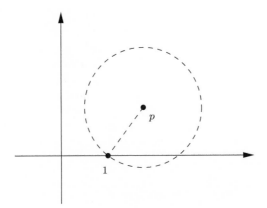

Figure 4.1. Geometric series based at p.

convergence. Doing so to (3) for $|w| < 1$ yields

(6) $\qquad \tan^{-1}(z) = \int_0^z \frac{dw}{1 + w^2} = \int_0^z \sum_{n=0}^{\infty} (-1)^n w^{2n} dw = \sum_{n=0}^{\infty} (-1)^n \frac{z^{2n+1}}{2n + 1}.$

Example 1.4. The following identity holds for $|z| < 1$ and $d \in \mathbf{N}$:

(7) $\qquad \frac{d!}{(1 - z)^{d+1}} = \sum_{n=0}^{\infty} (n + d)(n + d - 1)...(n + 1)z^n.$

This result is obtained by repeatedly differentiating the geometric series and changing the index of summation.

Example 1.4 leads to a useful general result.

Theorem 1.1. *Let $p(n)$ be a polynomial of degree d. Then $\sum_{n=0}^{\infty} p(n)z^n$ is a polynomial q in $\frac{1}{1-z}$ of degree $d + 1$ with no constant term.*

Proof. We induct on the degree of p. For $d = 0$ and thus $p(n) = c \neq 0$, we have the formula

$$\sum_{n=0}^{\infty} p(n)z^n = \frac{c}{1 - z}.$$

Hence $q(w) = cw$, and q has degree one, with no constant term. For the induction step, suppose we know the result for all polynomials of degree d. Consider a polynomial p of degree $d + 1$; we first write $p(n) = cn^{d+1} + r_d(n)$, where $c \neq 0$ and the degree of r_d is at most d. We can also write

$$n^{d+1} = (n + d + 1)(n + d)...(n + 1) - g_d(n),$$

where g_d is of degree d. Thus, we further have

(8) $\qquad p(n) = c(n + d + 1)(n + d)...(n + 1) - h_d(n),$

where h_d is of degree at most d. By induction and (8), we have reduced the matter to the special case when p is the polynomial $(n + d + 1)(n + d)...(n + 1)$; we must show that

$$\sum (n+d+1)(n+d)...(n+1)z^n$$

is a polynomial of degree $d+2$ in $\frac{1}{1-z}$ with no constant term. This conclusion follows by differentiating the geometric series $d+1$ times as in Example 1.4. □

Example 1.5. Let $p(n) = n^2$. First write $n^2 = (n+2)(n+1) - 3(n+1) + 1$. We then have, for $|z| < 1$,

$$\sum_{n=0}^{\infty} n^2 z^n = \sum_{n=0}^{\infty}(n+2)(n+1)z^n - \sum_{n=0}^{\infty} 3(n+1)z^n + \sum_{n=0}^{\infty} z^n$$

$$= \frac{2}{(1-z)^3} - \frac{3}{(1-z)^2} + \frac{1}{1-z}.$$

▶ **Exercise 4.1.** For $d = 1, 3, 4$, respectively, let $p(n) = n^d$. In each case find the corresponding polynomial q in $\frac{1}{1-z}$ from Theorem 1.1 such that

$$\sum_{n=0}^{\infty} p(n)z^n = q(\frac{1}{1-z}).$$

▶ **Exercise 4.2** (For those who know linear algebra). The mapping $p \to q$ from Theorem 1.1 is an invertible linear transformation between two vector spaces of polynomials. Find its matrix with respect to the usual bases.

▶ **Exercise 4.3.** In each case find an explicit formula (as a rational function) for the series and state where the formula is valid:

$$\sum_{n=0}^{\infty}(3+4i)^n z^n,$$

$$\sum_{n=0}^{\infty}(n^3 - 1)z^n.$$

2. The radius of convergence

Let $\sum_{n=0}^{\infty} a_n z^n$ be a given power series. By definition, this series converges at z if the limit $\lim_{N\to\infty} S_N(z)$ of the partial sums $S_N(z) = \sum_{n=0}^{N} a_n z^n$ exists. Let T denote the set of z for which this limit exists. For $z \in T$ we can define a function f by decreeing that $f(z)$ is the value of this limit. We saw in the previous section that a geometric series converged inside a circle and diverged outside this circle. We soon establish the corresponding elementary but fundamental result for arbitrary convergent power series.

First we make a simple observation. If a series converges, then the sequence of its individual terms must tend to 0. Since a convergent sequence is bounded, the terms of any convergent series must be bounded. For a general series, that test does little good. For a power series $\sum a_n z^n$, however, boundedness of the terms $a_n z^n$ will force convergence at any ζ such that $|\zeta| < |z|$. The next theorem makes this point precise and shows why power series (in one variable) define nice functions inside the *circle of convergence*.

Theorem 2.1. *Given a power series $\sum_{n=0}^{\infty} a_n z^n$, there is a nonnegative real number R (or the value infinity), called the radius of convergence of the series, such that the following hold:*

 1) *For $|z| > R$, the series diverges.*

 2) *For each r such that $0 \leq r < R$, the series converges absolutely and uniformly for $|z| \leq r$.*

 3) *The number R (or infinity) has the following value:*

(9) $$R = \sup\{r : |a_n| r^n \text{ is a bounded sequence}\}.$$

Proof. Define R as in (9), where *sup* means least upper bound. If $|z| > R$, then the terms of $\sum_{n=0}^{\infty} a_n z^n$ are unbounded and the series diverges. Therefore 1) holds. Now suppose that $0 \leq \rho < r < R$. Assume that $|a_n r^n| \leq M$. We claim that $\sum_n |a_n z^n|$ converges for $|z| \leq \rho$. The proof uses the comparison test:

(10) $$\sum_n |a_n z^n| \leq \sum_n |a_n| \rho^n = \sum_n |a_n| r^n (\frac{\rho}{r})^n \leq M \sum_n (\frac{\rho}{r})^n.$$

The geometric series on the far right-hand side of (10) converges since $|\frac{\rho}{r}| < 1$. Thus 2) and 3) also hold. $\qquad\square$

Remark 2.1. The number R of Theorem 2.1 can also be computed by Hadamard's *root test*. The condition $|a_n r^n| \leq M$ can be rewritten

(11) $$|r| \leq \frac{M^{\frac{1}{n}}}{|a_n|^{\frac{1}{n}}}.$$

Put $L = \lim \sup(|a_n|^{\frac{1}{n}})$. Let n tend to infinity in (11), and we get $|r| \leq \frac{1}{L}$. It follows that $\frac{1}{R}$ equals $\lim \sup(|a_n|^{\frac{1}{n}})$.

See Chapter 1 for the definition of *lim sup* and see [20] for more information. For the purpose of gaining intuition, we note that the lim sup of a sequence is the supremum of the set of its limit points. Furthermore if the limit of a sequence exists, then the limit and the lim sup are equal.

Example 2.1. The series $\sum \frac{z^n}{n!}$ converges for all z; thus $R = \infty$. The series $\sum (1+i)^n z^n$ converges for $|z| < \frac{\sqrt{2}}{2}$; thus $R = \frac{\sqrt{2}}{2}$. The series $\sum n! z^n$ converges only for $z = 0$; thus $R = 0$.

We can also consider series of the form $\sum a_n (z - p)^n$. We call p the *base point* and say that such a series is *based* at p, or expanded *about* p. In this case we obtain $\{z : |z - p| < R\}$ for the region of convergence of the series.

Theorem 2.1 says nothing about what happens on the circle $|z - p| = R$. Other than providing a few examples, we will leave this matter to more advanced texts.

Example 2.2. It is evident that the series $\sum (z - p)^n$ converges for $|z - p| < 1$ and diverges for $|z - p| > 1$. Thus $R = 1$. The series also diverges at z for $|z - p| = 1$. On the other hand, we have $R = 1$ also for the series $\sum \frac{(z-p)^n}{n^2}$, but now it converges everywhere on the circle $|z - p| = 1$.

▶ **Exercise 4.4.** Show that $\sum n z^n$ diverges if $|z| = 1$.

▶ **Exercise 4.5.** (Difficult) Show that $\sum \frac{z^n}{n}$ diverges if $z = 1$ but otherwise converges if $|z| = 1$.

▶ **Exercise 4.6.** Find the power series expansion for e^z about a general point z_0. Hint: Substitute $z - z_0$ in the known series and use the functional equation.

▶ **Exercise 4.7.** Find the power series expansion for $\frac{z}{z^4+9}$ about 0. Where does it converge?

▶ **Exercise 4.8.** Where does the series for $\frac{1}{1-z}$, expanded about the point $5i$, converge?

▶ **Exercise 4.9.** Find a formula for $\sum_{n=0}^{\infty} \frac{z^n}{4^{n+2}}$.

▶ **Exercise 4.10.** Find the radius of convergence of the following series:

$$\sum_{n=1}^{\infty} \frac{z^n n^n}{n!},$$

$$\sum_{n=1}^{\infty} z^{n!}.$$

▶ **Exercise 4.11.** Find the first 6 terms of the power series about 0 for $\frac{z}{e^z-1}$. See Section 5 and the Internet for more about *Bernoulli numbers*.

▶ **Exercise 4.12.** In solving this problem, assume by Corollary 5.1 of Chapter 2 that you may integrate or differentiate a power series term by term within its region of convergence.

1) In Example 1.3 we found the power series for $\tan^{-1}(z)$. Use it to find a series representation for $\frac{\pi}{6}$.

2) Find the power series for $\log(1 + z)$ and determine where it is valid.

▶ **Exercise 4.13.** Find the power series expansion for $\frac{1}{z}$ about a general nonzero point p and determine where it converges.

▶ **Exercise 4.14.** Expand $\frac{1}{1-z}$ in a series in negative powers of z. Where is this series valid?

▶ **Exercise 4.15.** Suppose that $f(z) = c_0 + c_1 z + ...$, where $c_0 \neq 0$ and the series converges near 0. Find the first two terms in the series for $\frac{1}{f(z)}$ at 0.

▶ **Exercise 4.16.** Expand $\frac{1}{(1-z)(2-z)}$ into a series that includes both positive and negative powers of z. Use partial fractions. Make sure that your series converges in $1 < |z| < 2$.

▶ **Exercise 4.17.** Find a simple expression for $\sum_{n=1}^{\infty}(-1)^n(n+1)(z-1)^n$ that is valid if $|z - 1| < 1$.

▶ **Exercise 4.18.** (Difficult) The ratio test fails when $\lim |\frac{a_{n+1}}{a_n}| = 1$. Prove the following generalization dealing with that case. Suppose that the limit of

$$n(1 - |\frac{a_{n+1}}{a_n}|)$$

exists and is larger than 1. Prove that $\sum |a_n|$ converges. Intuitively, when $|\frac{a_{n+1}}{a_n}|$ behaves like $1 - \frac{p}{n}$ for large n where $p > 1$, then the series converges. Relate this discussion to the series $\sum \frac{1}{n^p}$.

3. Generating functions

Let $\{a_n\}$, for $n \geq 0$, be a sequence of complex numbers. When we wish to consider the entire sequence as one entity, we can do so using *generating functions*.

Definition 3.1. The *ordinary generating function* of the sequence $\{a_n\}$ is the formal power series

$$\sum_{n=0}^{\infty} a_n z^n.$$

The *exponential generating function* of the sequence $\{a_n\}$ is the formal power series

$$\sum_{n=0}^{\infty} \frac{a_n z^n}{n!}.$$

In some circumstances we can use generating functions to answer interesting questions about a given sequence. For example, in Section 4 we find an explicit formula for the generating function of the sequence of Fibonacci numbers and use it to illuminate the golden ratio. Combinatorial analysis also offers many examples where one can use algebraic properties of generating functions to count the number of ways to do something interesting.

Choose a number x in $(0, 1)$. We can regard x as the generator of a sequence; we expand x as a decimal and consider the sequence of digits. Conversely, the sequence of digits determines x. Generating functions generalize this correspondence. A function with a convergent power series generates a sequence of complex numbers, and this sequence determines the function.

Remark 3.1. We first make a simple statement and then we say something harder. Both statements have profound consequences. Suppose we are given a convergent power series $f(z) = \sum c_n z^n$ and f is expressed in some explicit fashion. We can then recover the coefficients by *differentiation*; $c_k = \frac{f^{(k)}(0)}{k!}$. Corollary 4.1 of the Cauchy integral formula from Chapter 6 enables us to recover the coefficients by *integration*.

We give several examples of generating functions. If a_n is the constant sequence where $a_n = A$ for all n, then $\sum_{n=0}^{\infty} a_n z^n = A\frac{1}{1-z}$ as long as $|z| < 1$, and hence $A\frac{1}{1-z}$ is the ordinary generating function of the constant sequence $a_n = A$. Furthermore $\sum_{n=0}^{\infty} \frac{a_n z^n}{n!} = Ae^z$, and hence Ae^z is the exponential generating function of the same sequence. Let $a_n = p(n)$, where p is a polynomial of degree d. Then the ordinary generating function of this sequence is the polynomial q of degree $d + 1$ in the variable $\frac{1}{1-z}$, given by Theorem 1.1.

Example 3.1. Let $a_n = n^2$. For $|z| < 1$ we saw in Example 1.5 that

$$(12) \qquad \sum_{n=0}^{\infty} n^2 z^n = \sum_{n=0}^{\infty} (n+2)(n+1) z^n - \sum_{n=0}^{\infty} 3(n+1) z^n + \sum_{n=0}^{\infty} z^n$$

$$= \frac{2}{(1-z)^3} - \frac{3}{(1-z)^2} + \frac{1}{1-z} = \frac{z^2 + z}{(1-z)^3} = G(z).$$

Hence G is the ordinary generating function for the sequence of squares.

Assume we know the exponential generating function f of a sequence $\{a_n\}$ and its series converges in some neighborhood of 0. By Remark 3.1 we can recover the sequence: $a_n = (\frac{\partial}{\partial z})^n f(z)$ evaluated at $z = 0$. Even when the series does not converge, we can still regard a_n as a formal n-th derivative of f at 0.

▶ **Exercise 4.19.** A player flips a (biased) coin which lands heads up with probability p. The player wins as in classical tennis scoring. In other words, the winning possibilities are $(6, 0)$, $(6, 1)$, $(6, 2)$, $(6, 3)$, $(6, 4)$, $(7, 5)$, $(8, 6)$, and so on, where we write the number of heads first. Thus the wins are $(6, k)$ for $0 \leq k \leq 4$ and $(n + 2, n)$ for $n \geq 5$. Determine, as a function of p, the probability $f(p)$ of winning. Suggestion: First do the cases $(6, k)$ for $0 \leq k \leq 4$ separately. Then figure out the probability of winning from a $(5, 5)$ tie.

▶ **Exercise 4.20.** Consider the series for $\frac{1}{\cosh(z)}$. For $|z| < \frac{\pi}{2}$ it can be written (in terms of its exponential generating function) as

$$\sum_{n=0}^{\infty} \frac{z^n E_n}{n!}.$$

Find E_n for $0 \leq n \leq 6$.

▶ **Exercise 4.21.** For each $d \in \mathbf{N}$, show that $\sum_{n=1}^{\infty} n^d \frac{1}{2^n}$ is an integer.

▶ **Exercise 4.22.** Suppose we want a generating function for $n!$. Which type should we use? Why?

▶ **Exercise 4.23.** Find the ordinary generating function for the sequence $\{n^3\}$.

▶ **Exercise 4.24.** Put $a_0 = 0$ and $a_n = \frac{1}{n}$ for $n \geq 1$. Find the ordinary generating function for this sequence.

4. Fibonacci numbers

The Fibonacci numbers arise in many contexts in recreational mathematics. One can prove seemingly arbitrarily many nice things about them and the journal *The Fibonacci Quarterly* is devoted nearly entirely to them. These numbers are so much fun that we spend some effort playing with them and then explaining some of the ideas in deeper settings.

No one knows when these numbers were first noted, but they go back at least to 1202, in the work of Leonardo Fibonacci. These numbers are defined via a second-order recurrence relation. Put $F_0 = F_1 = 1$. For $n \geq 0$, put

$$(13) \qquad\qquad F_{n+2} = F_{n+1} + F_n.$$

The positive integer F_n is called the n-th Fibonacci number. The first few Fibonacci numbers are $1, 1, 2, 3, 5, 8, 13, 21, 34, 55, 89, 144$. A famous result is

$$\lim_{n\to\infty} \frac{F_{n+1}}{F_n} = \frac{1 + \sqrt{5}}{2} = \phi.$$

This limiting value ϕ is known as the *golden ratio*. Theorem 4.1 gives a formula for F_n, called Binet's formula, but which was known long before Binet (1843).

Given a sequence, we naturally consider its ordinary generating function. For the Fibonacci numbers this generating function is the formal power series

$$\sum_{n=0}^{\infty} F_n z^n.$$

This series converges near 0 to the explicit rational function f given in (14).

Proposition 4.1. *For $|z| < \frac{\sqrt{5}-1}{2}$,*

$$(14) \qquad \sum_{n=0}^{\infty} F_n z^n = \frac{1}{1 - z - z^2}.$$

Proof. Let $f(z)$ denote the sum. We then immediately obtain

$$(15) \qquad f(z) - 1 = \sum_{n=0}^{\infty} F_{n+1} z^{n+1}$$

and therefore

$$(16) \qquad f(z) - 1 - z = \sum_{n=0}^{\infty} F_{n+2} z^{n+2}.$$

Using (13), (15), and (16), we obtain

$$(17) \quad z^2 f(z) + z\left(f(z) - 1\right) = \sum_{n=0}^{\infty} (F_n + F_{n+1}) z^{n+2} = \sum_{n=0}^{\infty} F_{n+2} z^{n+2} = f(z) - 1 - z.$$

From (17) we obtain

$$f(z)(z^2 + z - 1) - z = -1 - z$$

from which formula (14) follows.

Where does the series converge? One can answer this question by expressing (14) in partial fractions. Alternatively, by Theorem 4.3 of Chapter 6, the series defined by (14) converges in the largest disk about 0 where the function is complex analytic. Since the expression $\frac{1}{1-z-z^2}$ is rational in z, it defines a complex analytic function away from the zero-set of the denominator. By the quadratic formula, the two singularities are at $\frac{-1\pm\sqrt{5}}{2}$. The singularity closer to 0 is $\frac{-1+\sqrt{5}}{2}$. Hence the series for f converges when $|z| < \frac{-1+\sqrt{5}}{2}$. $\qquad \square$

In order to derive Binet's explicit formula for F_n, we introduce a standard technique for solving constant coefficient linear recurrences. Suppose we are given constants $b_0, ..., b_{k-1}$ and the first k numbers $G_0, G_1, ..., G_{k-1}$ of a sequence. Assume that each subsequent term is determined by the rule

$$(18) \qquad G_{n+k} = \sum_{j=0}^{k-1} b_j G_{n+j}.$$

This recursive formula determines G_m for all m, and thus it defines a sequence of complex numbers. The integer k is called the *order* of the recurrence relation. The Fibonacci numbers are defined via the second-order recurrence (13). Let us recall a technique for solving such recurrences. We make the guess that there is a complex number λ such that $G_m = \lambda^m$ for all m. Next we plug this guess into

(18). After assuming that $\lambda \neq 0$ and dividing by λ^n, we obtain an equation for λ, called the characteristic equation:

$$(19) \qquad \lambda^k = \sum_{j=0}^{k-1} b_j \lambda^j.$$

We can rewrite (19) as $p(\lambda) = 0$, where

$$p(z) = z^k - \sum_{j=0}^{k-1} b_j z^j.$$

In the generic situation the polynomial equation $p(\lambda) = 0$ will have k distinct complex roots $r_1, \ldots, r_k$. We then put

$$G_n = \sum_{j=1}^{k} c_j r_j^n.$$

The c_j are unknown coefficients, which we can find by using the given k initial values to obtain a system of linear equations for them. One then easily checks that the method gives a valid solution. We will carry out this procedure for the Fibonacci numbers momentarily. We first mention that a slightly modified version of the procedure works even when the *characteristic equation* has repeated roots. Students might be familiar with the analogous issue when solving constant coefficient second-order differential equations.

Theorem 4.1 (Binet's formula). *The n-th Fibonacci number satisfies the following:*

$$F_n = \frac{1}{\sqrt{5}} \left(\frac{1 + \sqrt{5}}{2} \right)^{n+1} - \frac{1}{\sqrt{5}} \left(\frac{1 - \sqrt{5}}{2} \right)^{n+1}.$$

Proof. We follow the procedure outlined above. Assume $F_n = \lambda^n$. From the recurrence (13) we obtain the *characteristic equation* $\lambda^2 = \lambda + 1$. By the quadratic formula, the roots of this equation are $\frac{1 \pm \sqrt{5}}{2}$. For constants A, B to be determined we put

$$(20) \qquad F_n = A\left(\frac{1 + \sqrt{5}}{2}\right)^n + B\left(\frac{1 - \sqrt{5}}{2}\right)^n.$$

Using the two known values $F_0 = F_1 = 1$, we obtain a linear system of equations for A and B:

$$(21) \qquad 1 = A + B,$$

$$(22) \qquad 1 = A\frac{1 + \sqrt{5}}{2} + B\frac{1 - \sqrt{5}}{2}.$$

Solving this system yields $A = \frac{\sqrt{5}+1}{2\sqrt{5}}$ and $B = \frac{\sqrt{5}-1}{2\sqrt{5}}$. Plugging these values into (20) and simplifying gives the desired result. $\qquad \square$

Corollary 4.1. $\lim_{n \to \infty} \frac{F_{n+1}}{F_n} = \frac{1+\sqrt{5}}{2}$.

Proof. Write $\phi = \frac{1+\sqrt{5}}{2}$ and $\psi = \frac{1-\sqrt{5}}{2}$. By Binet's formula, we have

$$(23) \qquad \frac{F_{n+1}}{F_n} = \frac{\phi^{n+1} - \psi^{n+1}}{\phi^n - \psi^n}.$$

Note that $|\psi| < 1$ and hence $\lim_{n \to \infty} \psi^n = 0$. Therefore the limit of (23) as n tends to infinity is the golden ratio ϕ. $\qquad\qquad\square$

We wish to sketch two additional methods for establishing this corollary. First, we can appeal directly to the recurrence relation. Using it, we have

$$(24) \qquad \frac{F_{n+2}}{F_{n+1}} = \frac{F_{n+1} + F_n}{F_{n+1}} = 1 + \frac{F_n}{F_{n+1}}.$$

Assuming that the limit of the ratios is some nonzero number L, we let n tend to infinity in (24) to obtain the equation $L = 1 + \frac{1}{L}$, from which we obtain the characteristic equation $L^2 - L - 1 = 0$. Hence, if the limit exists, then its value is one of the roots of this equation, namely ϕ or ψ. But, $\psi < 0$, and hence the only possible value of the limit is ϕ. To prove that the limit exists, Exercise 4.26 suggests the following approach. Let $G_n = \frac{F_{n+1}}{F_n}$. Then the subsequences G_{2n} and G_{2n+1} are each bounded and monotone and hence have limits.

The second approach to the limit of the ratios involves the ratio test. We found already that the generating function for the Fibonacci numbers is the rational function $\frac{1}{z^2 - z - 1}$. The power series therefore converges in the largest disk about the origin on which $z^2 - z - 1 \neq 0$; thus it converges for $|z| < |\psi|$. On the other hand, the ratio test for convergence shows that the limit of the ratios of successive terms, if it exists, must be the reciprocal of this number, namely ϕ.

▶ **Exercise 4.25.** Use partial fractions to verify (14) and to determine where the series converges.

▶ **Exercise 4.26.** Fill in the details of the above methods for proving Corollary 4.1.

▶ **Exercise 4.27.** Where does the power series, based at 0, for $\frac{1}{1-(a+b)z+abz^2}$ converge?

5. An application of power series

In this section we use the exponential function in a stunning fashion to solve a natural problem whose statement is easy but whose solution requires cleverness.

We begin with the vague notion of the *shape* of a positive integer. We call k a *square number* if we can arrange k dots in a square, in other words, if $k = n^2$ for some n. For a similar geometric idea, we call m a *triangular number* if we can write m as the sum of the numbers from 1 to n for some n. It is well known that

$$(25) \qquad 1 + 2 + 3 + \ldots + n = \frac{n(n+1)}{2} = \frac{1}{2}n^2 + \frac{1}{2}n.$$

Gauss seems to have known this formula as a six-year-old. Indeed one can easily derive (25) as follows:

$$1 + 2 + 3 + \ldots + n = S,$$
$$n + (n-1) + \ldots + 1 = S.$$

Adding the two equations gives

$$n(n+1) = (1+n) + (1+n) + ... + (1+n) = 2S.$$

In other words, n copies of $n+1$ make $2S$ and thus $S = \frac{n(n+1)}{2}$.

▶ **Exercise 4.28.** Draw a suggestive picture explaining why (25) holds. Next find (with proof) the sum $1 + 3 + ... + (2n - 1)$. Draw another suggestive picture!

The formulas for the sum of the squares and the sum of the cubes provide further examples of explicit sums, usually stated and then proved by induction:

(26) $$1 + 4 + 9 + ... + n^2 = \frac{1}{3}n^3 + \frac{1}{2}n^2 + \frac{1}{6}n,$$

(27) $$1 + 8 + 27 + ... + n^3 = \frac{1}{4}n^4 + \frac{1}{2}n^3 + \frac{1}{4}n^2.$$

Given a positive integer p, we naturally seek a formula for the sum of the first n p-th powers. For each p the correct formula, when given, can be verified by induction. How can we *derive* the formula? Here is a straightforward but inefficient method. First we assume that the answer is a polynomial of degree $p+1$ in n, where we treat the coefficients as unknowns. Then we find the first $p+1$ values by hand. Finally we solve the resulting system of linear equations for the unknown coefficients.

This work can be done for all p at the same time, by combining our knowledge about the geometric and exponential series. We will find a formula for $\sum_{j=0}^{N-1} j^p$ using only the following ideas:

- The finite geometric series.
- The power series for e^z.
- The Leibniz rule (28) for the p-th derivative of a product:

(28) $$(\frac{d}{dz})^p(fg) = \sum_{k=0}^{p} \binom{p}{k} f^{(k)} g^{(p-k)}.$$

This rule follows by induction on p and the product rule $(fg)' = f'g + fg'$. See Exercise 4.30.

- The definition (29) of the Bernoulli numbers B_n:

(29) $$\frac{z}{e^z - 1} = \sum_{n=0}^{\infty} \frac{B_n}{n!} z^n.$$

Theorem 5.1.

$$\sum_{j=0}^{N-1} j^p = \sum_{k=0}^{p} \binom{p+1}{k+1} B_{p-k} \frac{N^{k+1}}{p+1}.$$

Proof. By the finite geometric series,

(30) $$\sum_{j=0}^{N-1} e^{jz} = \frac{e^{Nz} - 1}{e^z - 1} = \frac{e^{Nz} - 1}{z} \frac{z}{e^z - 1} = f(z)g(z).$$

We define f and g to be the factors in (30). We divided and multiplied by z in order to ensure that f and g have valid series expansions near 0. Taking p derivatives, evaluating at 0, and using the Leibniz rule, we get

$$(31) \qquad \sum_{j=0}^{N-1} j^p = (\frac{d}{dz})^p (f(z)g(z))(0) = \sum_{k=0}^{p} \binom{p}{k} f^{(k)}(0) g^{(p-k)}(0).$$

By Exercise 4.32, $f^{(k)}(0) = \frac{N^{k+1}}{k+1}$, and $g^{(p-k)}(0) = B_{p-k}$, by definition of the Bernoulli numbers. We conclude that

$$(32) \qquad \sum_{j=0}^{N-1} j^p = \sum_{k=0}^{p} \binom{p}{k} B_{p-k} \frac{N^{k+1}}{k+1} = \sum_{k=0}^{p} \binom{p+1}{k+1} B_{p-k} \frac{N^{k+1}}{p+1}.$$

$\square$

For example, we can now derive formulas such as

$$(33) \qquad \sum_{j=1}^{n} j^{10} = \frac{1}{11} n^{11} + \frac{1}{2} n^{10} + \frac{5}{6} n^9 - n^7 + n^5 - \frac{1}{2} n^3 + \frac{5}{66} n.$$

▶ **Exercise 4.29.** Verify (33).

▶ **Exercise 4.30.** Verify the Leibniz rule by induction.

▶ **Exercise 4.31.** The following identity was used in (32):

$$\frac{1}{k+1} \binom{p}{k} = \frac{1}{p+1} \binom{p+1}{k+1}.$$

Verify it algebraically. Also give a combinatorial proof. Suggestion: Count (using two different approaches) how many ways one can select a team of $k+1$ people from a collection of $p+1$ people and designate one of them captain.

▶ **Exercise 4.32.** Verify that $f^{(k)}(0) = \frac{N^{k+1}}{k+1}$, where $f(z) = \frac{e^{Nz}-1}{z}$ as above.

▶ **Exercise 4.33.** Determine the first six Bernoulli numbers.

▶ **Exercise 4.34.** Find $\sum_{j=1}^{n} j^8$ by the method of undetermined coefficients.

6. Rationality

We give a criterion for deciding whether a power series defines a rational function, after first recalling when a decimal expansion represents a rational number.

Given the decimal expansion of a real number x, it is possible to decide whether x is rational. A necessary and sufficient condition is that the decimal expansion eventually repeats. Let us be more precise. We may ignore the integer part of x without loss of generality; hence we assume $0 \le x < 1$ and write

$$x = \sum_{n=1}^{\infty} a_n (\frac{1}{10})^n = .a_1 a_2 a_3 ....$$

Then x is rational if and only if there are two natural numbers N and K with the following property. For $n \ge N$, and for all m, we have

$$a_{n+Km} = a_n.$$

In other words, the decimal expansion looks as follows:

$$x = .a_1....a_N a_{N+1}...a_{N+K-1} a_N a_{N+1}...a_{N+K-1} a_N a_{N+1}...a_{N+K-1}....$$

The point is that the string of K digits from a_N to a_{N+K-1} repeats forever. The case $K = 1$ is possible; it occurs when the decimal expansion is eventually constant. A terminating expansion occurs when $a_j = 0$ for $j \geq N$.

▶ **Exercise 4.35.** Express the decimal .198198198... as a rational number. Justify your method, using the definition of convergent infinite series.

▶ **Exercise 4.36.** Do the same for the decimal .12345198198198....

▶ **Exercise 4.37.** Prove both implications in the above test for rationality.

One of the remarkable things about power series is that a similar test works for deciding whether a power series represents a rational function. A rational function is a function that can be expressed as the ratio of two polynomials. We state and prove this test after introducing some notation.

Let f be a complex analytic function, defined near 0 in $\mathbf{C}$. In the next chapter we will prove that f has a convergent power series expansion near the origin. In other words there is a positive number R such that we can write

$$f(z) = \sum_{n=0}^{\infty} a_n z^n,$$

where the series converges for $|z| < R$. Conversely, such a series defines a complex analytic function in the disk determined by $|z| < R$. For our present purpose we are given a convergent series, and we wish to know whether it represents a rational function. We consider the Taylor polynomials and remainder terms for f:

(34)
$$j_k(z) = \sum_{n=0}^{k} a_n z^n,$$

(35)
$$R_k(z) = f(z) - j_k(z) = \sum_{n=k+1}^{\infty} a_n z^n.$$

Notice that $R_k(z)$ is divisible by z^{k+1}.

Let us now consider the infinite array defined as follows. The 0-th row is $f(z)$. For $k \geq 1$, we let the k-th row be the function

$$\frac{f(z) - j_{k-1}(z)}{z^k} = \frac{R_{k-1}(z)}{z^k}.$$

For example the first three rows of the array are given in (36):

(36)
$$\begin{pmatrix} f(z) \\[6pt] \frac{f(z)-f(0)}{z} \\[6pt] \frac{f(z)-f(0)-f'(0)z}{z^2} \\[2pt] ... \end{pmatrix}.$$

If we replace each row in (36) with the corresponding list of coefficients, we get the infinite array whose j, k-entry is a_{j+k}:

$$(37) \qquad \begin{pmatrix} a_0 & a_1 & a_2 & a_3 & ... \\ a_1 & a_2 & a_3 & ... & \\ a_2 & a_3 & a_4 & ... & \\ ... & & & & \end{pmatrix}.$$

A matrix such as (37) is called a Hankel matrix.

The test for rationality, in the language of linear algebra, is that this matrix has finite rank. In concrete terms, the test states that f is rational if and only if there is an integer N such that, whenever $n \geq N$, the n-th row is a linear combination of the first N rows. We naturally wish to express this condition directly in terms of f.

Suppose for some N that

$$(38) \qquad \frac{R_N(z)}{z^{N+1}} = \sum_{k=0}^{N-1} c_k \frac{R_k(z)}{z^{k+1}}.$$

Equivalently we are supposing that the N-th row of (37) is a linear combination of the previous rows. Clearing denominators in (38), we obtain

$$(39) \quad f(z) - j_N(z) = R_N(z) = \sum_{k=0}^{N-1} c_k R_k(z) z^{N-k} = \sum_{k=0}^{N-1} c_k \left(f(z) - j_k(z) \right) z^{N-k}.$$

Formula (39) is an affine equation for $f(z)$. We solve it by gathering all the $f(z)$ terms on one side, to get

$$f(z) - f(z) \sum_{k=0}^{N-1} c_k z^{N-k} = j_N(z) - \sum_{k=0}^{N-1} c_k j_k(z) z^{N-k}.$$

Factoring out $f(z)$ and dividing gives us an explicit formula for f as a rational function:

$$(40) \qquad f(z) = \frac{j_N(z) - \sum_{k=0}^{N-1} c_k j_k(z) z^{N-k}}{1 - \sum_{k=0}^{N-1} c_k z^{N-k}}.$$

This explicit formula (40) shows that the degrees of the numerator and denominator are at most N. It follows that all subsequent rows in the Hankel matrix (37) are linear combinations of the first N rows as well.

Conversely, suppose f is rational. Then one obtains constants c_k by setting the denominator of (40) equal to the denominator of f. One can then verify (38). We summarize this discussion by the following result:

Theorem 6.1. *Suppose $\sum_{n=0}^{\infty} a_n z^n$ converges near $z = 0$ to $f(z)$. The following are equivalent:*

1) *f is a rational function.*

2) *The infinite matrix (37) has finite rank.*

3) *There is an N such that $\frac{R_N(z)}{z^N}$ is a linear combination of the $\frac{R_k(z)}{z^k}$ for $k < N$ as in (38).*

▶ **Exercise 4.38.** Use the proof of Theorem 6.1 to show that $\sum_{n=0}^{\infty} z^n$ defines the rational function $\frac{1}{1-z}$ near $z = 0$.

▶ **Exercise 4.39.** Use the proof of Theorem 6.1 to show that $\sum_{n=0}^{\infty} n z^n$ defines a rational function near $z = 0$.

▶ **Exercise 4.40.** If p is a polynomial, show by this method that $\sum_{n=0}^{\infty} p(n) z^n$ defines a rational function near $z = 0$.

▶ **Exercise 4.41.** Consider a power series whose coefficients repeat the pattern $1, 1, 0, -1, -1$. Thus the series starts

$$1 + z - z^3 - z^4 + z^5 + z^6 - z^8 - z^9 \ldots$$

Show that this series defines a rational function f and write f explicitly in lowest terms.

▶ **Exercise 4.42.** Show that $\sum z^{n!}$ does not represent a rational function.

Chapter 5

Complex Differentiation

We have so far often clarified ideas involving two real variables by expressing them in terms of one complex variable. Such insight will be particularly compelling in this chapter. We gave a provisional definition of complex analytic function in Chapter 2. We can think of a function defined on a subset of $\mathbf{C}$ as depending on both $x = \mathrm{Re}(z)$ and $y = \mathrm{Im}(z)$. In some sense, a complex analytic function is a function that depends on only the particular combination z given by $z = x + iy$. It is therefore independent of $\bar{z}$. The Cauchy-Riemann equations make everything transparent. We begin by stating the fundamental result from elementary complex analysis; three possible definitions of *complex analytic* function lead to the same notion. We postpone the proofs of these assertions until Chapter 6, after we have introduced complex integration and can develop the Cauchy theory.

1. Definitions of complex analytic function

Let Ω be an open subset of $\mathbf{C}$ and suppose $f : \Omega \to \mathbf{C}$ is a function. We give three possible candidates for the definition of complex analytic function.

Definition 1.1 (Convergent power series). The function f is complex analytic on Ω if the following holds: for all $p \in \Omega$, there is a disk about p, lying in Ω, on which f can be developed in a convergent power series:

$$(1) \qquad f(z) = \sum_{n=0}^{\infty} a_n (z - p)^n.$$

Definition 1.2 (Cauchy-Riemann equations). The function f is complex analytic on Ω if the following holds: f is continuously differentiable, and for all $p \in \Omega$,

$$(2) \qquad \frac{\partial f}{\partial \bar{z}}(p) = 0.$$

Thus f satisfies the partial differential equation $\frac{\partial f}{\partial \bar{z}} = 0$.

Definition 1.3 (Difference quotient). The function f is complex analytic on Ω if the following holds: for all $p \in \Omega$, f is approximately complex linear at p. In other words, f has a complex derivative $f'(p)$, defined by the existence of the limit in (3):

$$(3) \qquad\qquad f'(p) = \lim_{\zeta \to 0} \frac{f(p + \zeta) - f(p)}{\zeta}.$$

Definition 1.3 is appealing because it is the natural analogue for $\mathbf{C}$ of the difference quotient definition of differentiability for a function on $\mathbf{R}$. On the other hand, it is much harder for a function of a complex variable to be differentiable in the sense of (3) than it is for a function to be differentiable on the real line. Because ζ is complex in (3), the existence of the limit turns out to be a surprisingly restrictive assumption. For example, the infinitely smooth and immensely useful functions $z \to |z|^2$ and $z \to x = \frac{z + \overline{z}}{2}$ do not satisfy (3).

Remark 1.1. The existence of the limit (3) forces f to be continuous. It also forces the directional derivative to exist in each direction. The consequences of (3) are subtle. Consider the following example where (3) fails. For $z \neq 0$, put $f(z) = \frac{\text{Im}(z^2)}{|z|^2}$ and put $f(0) = 0$. Then f is not continuous at 0, although the directional derivative $\frac{\partial f}{\partial v}(0)$ exists in every direction v.

▶ **Exercise 5.1.** Show that the functions $|z|^2$ and $z + \overline{z}$ satisfy neither (2) nor (3). Show that $z^a \overline{z}^b$ satisfies (2) only if $b = 0$.

▶ **Exercise 5.2.** Verify the statements made in Remark 1.1.

▶ **Exercise 5.3.** Suppose a Hermitian symmetric polynomial Φ satisfies $\frac{\partial}{\partial \overline{z}}(\Phi) = 0$. Prove that Φ is a constant.

Definition 1.3 can be rewritten in an important manner which illustrates that differentiability means approximate linearity. We write

$$(4) \qquad\qquad f(p + \zeta) = f(p) + f'(p)\zeta + E(p, \zeta),$$

where the function E is defined by (4). If the limit in (3) exists, then E satisfies

$$(5) \qquad\qquad \lim_{\zeta \to 0} \frac{E(p, \zeta)}{\zeta} = 0.$$

We call E an error function. The infinitesimal behavior of f at p is multiplication by $f'(p)$, and thus $f'(p)$ is the complex analogue of slope. Recall from Chapter 1 that multiplication by a complex number is a special kind of linear transformation of $\mathbf{R}^2$, one whose matrix satisfies the Cauchy-Riemann equations. In the next section we will develop this idea in detail.

Definition 1.2 states informally that f is a nice function that happens to be *independent* of $\overline{z}$. Definition 1.1 seems stronger; a function satisfying (1) is infinitely differentiable and, whatever it means to say so, surely independent of $\overline{z}$. Definition 1.2 says that, at each point, f is infinitesimally independent of $\overline{z}$. It then follows that f really is a nice function and it is independent of $\overline{z}$. Definition 1.3 makes the weakest assumptions on f, but all three definitions turn out to be equivalent.

2. Complex differentiation

We start with an intuitive discussion of differential forms. In Section 5 we will give a more precise treatment. For the moment, we take dx and dy as understood objects. We define dz and $d\bar{z}$ by $dz = dx + i\,dy$ and $d\bar{z} = dx - i\,dy$. It follows that $dx = \frac{dz+d\bar{z}}{2}$ and that $dy = \frac{dz-d\bar{z}}{2i}$. Let Ω be an open subset of $\mathbf{C}$. Recall that a function $P : \Omega \to \mathbf{C}$ is *smooth* if its partial derivatives of all orders exist and are themselves continuous. A differential 1-form ω on Ω is an expression of the form $\omega = Pdx + Qdy$, where P and Q are smooth functions on Ω. Since dx and dy are linear combinations of dz and $d\bar{z}$, we can also write ω as follows:

$$\omega = Pdx + Qdy = (\frac{P}{2} + \frac{Q}{2i})dz + (\frac{P}{2} - \frac{Q}{2i})d\bar{z} = Adz + Bd\bar{z}.$$

The *total differential df* of a smooth function provides an example of a differential 1-form. Let f be a smooth function of two variables. Presuming that the differentials dx and dy have been given a precise meaning, we put

$$(6) \qquad df = \frac{\partial f}{\partial x}dx + \frac{\partial f}{\partial y}dy.$$

Based on (6), we introduce differential operators $\frac{\partial}{\partial z}$ and $\frac{\partial}{\partial \bar{z}}$ in order to write

$$(7) \qquad df = \frac{\partial f}{\partial z}dz + \frac{\partial f}{\partial \bar{z}}d\bar{z}.$$

Since $dz = dx + idy$ and $d\bar{z} = dx - idy$, we set the two expressions for df equal. Doing so then forces the following fundamental definitions:

Definition 2.1.

$$(8) \qquad \frac{\partial}{\partial z} = \frac{1}{2}\left(\frac{\partial}{\partial x} - i\frac{\partial}{\partial y}\right),$$

$$(9) \qquad \frac{\partial}{\partial \bar{z}} = \frac{1}{2}\left(\frac{\partial}{\partial x} + i\frac{\partial}{\partial y}\right).$$

We also have the notion of differential 2-form. In the plane, the only examples are multiples of $dx \wedge dy$. The wedge product is a fascinating concept; we can multiply differential forms to obtain differential forms of higher type. We discuss the wedge product in Section 6. For now we note only the rules $dx \wedge dx = 0 = dy \wedge dy$ and $dx \wedge dy = -dy \wedge dx$. It follows that

$$(10) \qquad dx \wedge dy = \frac{i}{2}dz \wedge d\bar{z}.$$

▶ **Exercise 5.4.** Derive (8) and (9) by equating (6) and (7).

▶ **Exercise 5.5.** Assume $dz = dx + idy$ and $d\bar{z} = dx - idy$. Using the rules for wedge products, verify (10).

▶ **Exercise 5.6.** Express the operator $\frac{\partial^2}{\partial x^2} + \frac{\partial^2}{\partial y^2}$ in terms of z and $\bar{z}$ derivatives.

Speaking informally, we say that f is independent of $\bar{z}$ if and only if $\frac{\partial f}{\partial \bar{z}} = 0$.

Definition 2.2 (The Cauchy-Riemann operator). Let g be a continuously differentiable function on an open subset of $\mathbf{R}^2$. We define $\bar{\partial}g$ by

$$\bar{\partial}g = \frac{\partial g}{\partial\bar{z}}\,d\bar{z}.$$

Let $\omega = Pdx + Qdy$ be a 1-form; by definition ω is continuously differentiable if P and Q are. In this case we define $\bar{\partial}\omega$ by

$$(11) \qquad\qquad \bar{\partial}\omega = \frac{\partial P}{\partial\bar{z}}\,d\bar{z}\wedge dx + \frac{\partial Q}{\partial\bar{z}}\,d\bar{z}\wedge dy.$$

We express Definition 2.2 using complex derivatives. If $\omega = Adz + Bd\bar{z}$, then we obtain the simpler formula

$$(12) \qquad\qquad \bar{\partial}\omega = \frac{\partial A}{\partial\bar{z}}d\bar{z}\wedge dz.$$

▶ **Exercise 5.7.** Show that (11) and (12) are equivalent.

We summarize this section by stating the next result.

Proposition 2.1. *Suppose Ω is open and $f : \Omega \to \mathbf{C}$ is continuously differentiable. The following are equivalent:*

- *f is independent of $\bar{z}$.*
- *$\bar{\partial}f = 0$ on Ω.*
- *$df = \frac{\partial f}{\partial z}dz$ on Ω.*
- *$\frac{\partial f}{\partial x} + i\frac{\partial f}{\partial y} = 0$ on Ω.*

3. The Cauchy-Riemann equations

Following common practice, we will use the words *Cauchy-Riemann equations* in several similar contexts. We defined in Chapter 1, and we recall below, what it means to say that a two-by-two matrix satisfies the Cauchy-Riemann equations. We will say that a differentiable function $F : \mathbf{R}^2 \to \mathbf{R}^2$ satisfies the Cauchy-Riemann equations if its derivative $dF(p)$, regarded as a two-by-two matrix, satisfies these equations at each point. In this case, we may write $F = (u, v)$ or $f = u + iv$. We then sometimes will say that f and/or u, v satisfy these equations. By the end of the chapter we will say that f satisfies the Cauchy-Riemann equations if $\frac{\partial f}{\partial\bar{z}} = 0$.

We first recall the definition of the complex numbers using two-by-two matrices. We identify a complex number $a+ib$ with the linear transformation from $\mathbf{C}$ to itself given by multiplication by $a + ib$. If we consider this mapping instead as a linear transformation from $\mathbf{R}^2$ to itself, then the matrix of this linear mapping (with respect to the usual basis) has the form

$$(13) \qquad\qquad \begin{pmatrix} a & -b \\ b & a \end{pmatrix}.$$

Definition 3.1. A two-by-two matrix of real numbers satisfies the Cauchy-Riemann equations if it has the form (13).

A real linear mapping from $L : \mathbf{R}^2 \to \mathbf{R}^2$ defines a complex linear mapping from $\mathbf{C}$ to $\mathbf{C}$, namely multiplication by $a + ib$, if and only if the matrix of L is of the form (13). Thus the matrix of a real linear mapping L satisfies the Cauchy-Riemann equations if and only if L defines a *complex linear* map from $\mathbf{C}$ to $\mathbf{C}$.

Similarly, let $F : \mathbf{R}^2 \to \mathbf{R}^2$ be a differentiable function. We write $F = (u, v)$ in terms of its component functions. Its derivative $dF(z)$ at the point z is the best linear approximation to F at z. It is a linear transformation represented by the two-by-two matrix

$$(14) \qquad\qquad dF = \begin{pmatrix} \frac{\partial u}{\partial x} & \frac{\partial u}{\partial y} \\ \frac{\partial v}{\partial x} & \frac{\partial v}{\partial y} \end{pmatrix}.$$

Definition 3.2. A differentiable function satisfies the Cauchy-Riemann equations on an open set Ω if, for each $z \in \Omega$, the derivative matrix $dF(z)$ satisfies the Cauchy-Riemann equations.

Stated in terms of its real and imaginary parts, F satisfies the Cauchy-Riemann equations if and only if $\frac{\partial u}{\partial x} = \frac{\partial v}{\partial y}$ and $\frac{\partial u}{\partial y} = -\frac{\partial v}{\partial x}$ at each point. Modern complex analysis, especially in higher dimensions, has gained tremendously by writing these equations in terms of z and $\overline{z}$. For clarity and emphasis we repeat this key point.

Corollary 3.1. *A continuously differentiable complex-valued function $f = u + iv$ satisfies the Cauchy-Riemann equations $\overline{\partial} f = 0$ if and only if $\frac{\partial u}{\partial x} = \frac{\partial v}{\partial y}$ and $\frac{\partial u}{\partial y} = -\frac{\partial v}{\partial x}$.*

▶ **Exercise 5.8.** Put $F(x, y) = (x^2 - y^2, 2xy)$. Find $dF(x, y)$ and verify that F satisfies the Cauchy-Riemann equations. Write $F(x, y)$ as a function of z.

▶ **Exercise 5.9.** Put $F(x, y) = (x^3 - 3xy^2, 3x^2 y - y^3)$. Find $dF(x, y)$ and verify that F satisfies the Cauchy-Riemann equations. Write $F(x, y)$ as a function of z.

▶ **Exercise 5.10.** Put $F(x, y) = (e^x \cos(y), e^x \sin(y))$. Find $dF(x, y)$ and verify that F satisfies the Cauchy-Riemann equations. Write $F(x, y)$ as a function of z.

▶ **Exercise 5.11.** For $x, y > 0$, put $F(x, y) = (\frac{1}{2} \log(x^2 + y^2), \tan^{-1}(\frac{y}{x}))$. Find $dF(x, y)$ and verify that F satisfies the Cauchy-Riemann equations. Write $F(x, y)$ as a function of z.

We now make the needed connections between real and complex derivatives. Let g be a differentiable complex-valued function on an open set Ω in $\mathbf{R}^2$. By definition of the derivative, g is approximately linear. In other words, for each point $(x, y) \in \Omega$ and each sufficiently small vector (h, k) we can write

$$(15) \qquad g(x + h, y + k) = g(x, y) + \frac{\partial g}{\partial x}(x, y)h + \frac{\partial g}{\partial y}(x, y)k + E(x, y, h, k),$$

where the error term $E(x, y, h, k)$ is small in the sense that

$$\lim_{(h,k) \to (0,0)} \frac{E(x, y, h, k)}{\sqrt{h^2 + k^2}} = 0.$$

The differentiable function g is called *continuously differentiable* if the partial derivatives are themselves continuous functions. Formula (15) provides the first-order Taylor approximation of g. It is especially useful to rewrite this formula using partial derivatives with respect to z and $\overline{z}$. We obtain the following expression:

$$(16) \qquad g(z+\eta) = g(z) + \frac{\partial g}{\partial z}(z)\eta + \frac{\partial g}{\partial \overline{z}}(z)\overline{\eta} + E(z,\eta).$$

To see the equivalence of (15) and (16), we write $z = x + iy$, put $\eta = h + ik$, and use (8) and (9):

$$\frac{\partial g}{\partial z}\eta + \frac{\partial g}{\partial \overline{z}}\overline{\eta} = \frac{1}{2}\Big(\frac{\partial g}{\partial x} - i\frac{\partial g}{\partial y}\Big)(h+ik) + \frac{1}{2}\Big(\frac{\partial g}{\partial x} + i\frac{\partial g}{\partial y}\Big)(h-ik) = \frac{\partial g}{\partial x}h + \frac{\partial g}{\partial y}k.$$

In the notation of (16), the requirement for differentiability becomes

$$(17) \qquad \lim_{\eta \to 0} \frac{E(z,\eta)}{|\eta|} = 0.$$

The Cauchy-Riemann equations now lead to great simplification. The differentiable function g satisfies them on Ω if and only if $\frac{\partial g}{\partial \overline{z}} = 0$ on Ω. If g satisfies these equations, then we define $g'(z)$ to be $\frac{\partial g}{\partial z}(z)$. By (16) we see that

$$(18) \qquad g'(z) = \lim_{\zeta \to 0} \frac{g(z+\zeta) - g(z)}{\zeta}.$$

On the other hand, if the limit on the right-hand side of (18) exists, then g must satisfy the Cauchy-Riemann equations.

We summarize these calculations as follows.

Proposition 3.1. *Assume that g is continuously differentiable. Then $\frac{\partial g}{\partial \overline{z}} = 0$ on Ω if and only if the limit in (18) exists at each z in Ω.*

We sound one warning. The Cauchy-Riemann equations $\frac{\partial g}{\partial \overline{z}} = 0$ can hold at one point without the limit in (18) existing there.

Example 3.1. Put $g(z) = \frac{\overline{z}^2}{z}$ for $z \neq 0$ and put $g(0) = 0$. Then g satisfies the Cauchy-Riemann equations at 0, but the limit in (18) does not exist there.

▶ **Exercise 5.12.** Verify the statements in Example 3.1.

▶ **Exercise 5.13.** Determine whether there is a function satisfying $f(z+h) = f(z) + h^2$ for all z and h.

▶ **Exercise 5.14.** Prove some of the usual rules from calculus for derivatives in the complex setting.

- Show using (18) that $\frac{d}{dz}(z^n) = nz^{n-1}$ for $n \geq 0$.
- Show using (18) that $\frac{d}{dz}(\frac{1}{z}) = \frac{-1}{z^2}$ for $z \neq 0$.
- Assume f, g are complex differentiable. Show that fg also is and that $(fg)' = f'g + fg'$.
- Assume f, g are complex differentiable. Show that their composition also is and that $(g \circ f)'(z) = g'(f(z))f'(z)$.

An interesting technical point arises when considering the limit in (18) for complex functions. If we make no assumption about g other than that the limit in (18) exists at all z in Ω, then it follows that g is differentiable and, even more, that g is infinitely differentiable. Hence many authors define g to be complex analytic on Ω

when the limit in (18) exists at each point of Ω; in other words, the third definition from Section 1 is taken as the starting point. It is then possible to establish the main results on complex integration by using Goursat's proof of Cauchy's theorem. It is also possible to begin with the second definition from Section 1. Under the stronger assumption that a complex analytic function is continuously differentiable, we can apply Green's theorem directly to obtain Cauchy's theorem and hence all the main results. See also [**1, 17, 19**] for more discussion about this matter.

These considerations depend strongly upon working on an open set. If the limit in (18) exists at a single point, then we can say virtually nothing. The following standard examples illustrate some of the difficulties arising from working at only one point. We emphasize that all three of our possible definitions of complex analytic function insist that something be true at every point of an open set.

Example 3.2. Let f be an arbitrary bounded function on $\mathbf{C}$. In particular f could be discontinuous at every point. Define $g : \mathbf{C} \to \mathbf{C}$ by $g(z) = |z|^2 f(z)$. We claim that g is complex differentiable at the origin. To compute the limit in (18), we observe

$$(19) \qquad \frac{g(0 + \zeta) - g(0)}{\zeta} = \frac{|\zeta|^2 f(\zeta)}{\zeta} = \bar{\zeta} f(\zeta).$$

Since f is bounded, the limit as ζ tends to 0 in (19) exists and equals 0. On the other hand, although g is continuous at 0, it can be discontinuous at every other point in $\mathbf{C}$.

The next example is even worse!

Example 3.3. Consider a function f on $\mathbf{C}$ that is defined to be 0 on the axes but takes completely arbitrary real values elsewhere. Writing $f = u + iv$ gives $v = 0$ by definition. Hence $\frac{\partial v}{\partial x}(0) = 0$ and $\frac{\partial v}{\partial y}(0) = 0$. Since u is constant on the axes, we also see that $\frac{\partial u}{\partial x}(0) = 0$ and $\frac{\partial u}{\partial y}(0) = 0$. Hence the real form of the Cauchy-Riemann equations for f holds at 0, and yet f need not be continuous there.

4. Orthogonal trajectories and harmonic functions

Recall that the Euclidean dot (or inner) product of vectors α and β in $\mathbf{R}^2$ is given by

$$\alpha \cdot \beta = \langle \alpha, \beta \rangle = \alpha_1 \beta_1 + \alpha_2 \beta_2.$$

The length $||\alpha||$ of the vector α is of course $\sqrt{\alpha_1^2 + \alpha_2^2}$. Assume that α and β are based at the same point and that the angle between them, expressed in radians, is θ. The geometric meaning of the dot product arises from the famous relationship

$$\langle \alpha, \beta \rangle = \cos(\theta) ||\alpha|| \; ||\beta||.$$

In particular, two vectors are perpendicular if and only if their dot product is zero. Mathematicians often say *orthogonal* instead of perpendicular.

Let $u : \mathbf{R}^2 \to \mathbf{R}$ be a function. The *level set* $\{u = c\}$ is the set of (x, y) such that $u(x, y) = c$. Level sets play a fundamental role in mathematics, physics, economics, geology, and meteorology. The reader should try to think of examples from each of these subjects. The concept becomes more interesting when u is differentiable. In that case, for each point (x, y), the gradient vector $\nabla u(x, y)$ is perpendicular to the level set $u = u(x, y)$. We do not prove this statement, but it is proved in many calculus books. See for example [**24**].

Consider next two differentiable functions u and v on $\mathbf{R}^2$. Choose a point (x, y) and consider the level sets $\{u = u(x, y)\}$ and $\{v = v(x, y)\}$. They intersect at (x, y); they are perpendicular there if and only if $\nabla u(x, y)$ and $\nabla v(x, y)$ are perpendicular. In other words, these level sets are perpendicular at (x, y) if and only if

$$\frac{\partial u}{\partial x}\frac{\partial v}{\partial x} + \frac{\partial u}{\partial y}\frac{\partial v}{\partial y} = 0.$$

Suppose that $\nabla u(x, y) = (a, b)$. Then $\nabla u(x, y)$ is perpendicular to $\nabla v(x, y)$ if and only if $\nabla v(x, y)$ is a multiple of $(-b, a)$. Aha! If u and v satisfy the Cauchy-Riemann equations, then their level sets are orthogonal. We summarize these comments in the following fundamental fact.

Theorem 4.1. *Let $f = u + iv$ be complex analytic. At each point of intersection, the level sets of u and v intersect orthogonally.*

Example 4.1. Let $f(z) = z^2$. The level sets of u and v are the hyperbolas given by $x^2 - y^2 = c_1$ and $2xy = c_2$. This family of hyperbolas intersects itself orthogonally. See Figure 3.3.

▶ **Exercise 5.15.** Regard $z, w \in \mathbf{C}$ as vectors in $\mathbf{R}^2$. Show that $\langle z, w \rangle = \mathrm{Re}(z\overline{w})$. Draw an appropriate picture.

▶ **Exercise 5.16.** Sketch the level sets of the real and imaginary parts of z^3.

▶ **Exercise 5.17.** Let $f(z) = e^z$. What are the real and imaginary parts of f in terms of x, y? Graph their level sets; show that one gets orthogonal trajectories.

▶ **Exercise 5.18.** Describe and graph the level sets for the functions u and v if

$$f(x, y) = \frac{1}{2}\log(x^2 + y^2) + i\tan^{-1}(\frac{y}{x}) = u(x, y) + iv(x, y),$$

for $x, y > 0$. Determine whether f satisfies the Cauchy-Riemann equations.

5. A glimpse at harmonic functions

Harmonic functions play a fundamental role in pure and applied mathematics. Many equivalent definitions are possible, and we do not mention most of them. We give two simple possible definitions. Assuming that u is twice continuously differentiable, we say that u is harmonic if its Laplacian Δu is 0 (Definition 5.1). We show that u satisfies the mean value property (Definition 5.2). Conversely if we assume that u satisfies the mean value property and is twice differentiable, then we show that u is harmonic. Our proof uses Taylor's formula in two variables. We do not attempt to show that the mean value property guarantees that u is twice differentiable.

This section develops the connection between complex analytic functions and harmonic functions. This connection lies at the basis of many applications of complex analysis. The real and imaginary parts of a complex analytic function are harmonic and their level sets form orthogonal trajectories. For this reason complex analysis arises in various applications such as electrostatics and fluid flow. We mention that Exercise 5.19 should help with the theoretical issues.

Definition 5.1. Let Ω be an open subset of $\mathbf{C}$ and let $u : \Omega \to \mathbf{R}$ be a twice continuously differentiable function. Then u is *harmonic* if

$$(20) \qquad \Delta(u) = \frac{\partial^2 u}{\partial x^2} + \frac{\partial^2 u}{\partial y^2} = 0$$

at all points of Ω.

Exercise 5.6 asked for a formula for Δ in terms of derivatives with respect to z and $\overline{z}$. That formula is

$$(21) \qquad \Delta = 4 \frac{\partial}{\partial z} \frac{\partial}{\partial \overline{z}}.$$

Often formula (21) clearly explains crucial properties of harmonic functions and the connections with complex analysis. For example, (21) shows immediately why the real and imaginary parts of a complex analytic function are harmonic. We prove Theorem 5.1 below using real variable techniques, but we urge the reader in Exercise 5.19 to try to prove it using complex derivatives. The next definition gives a more geometric definition of harmonic function.

Definition 5.2. Let Ω be an open subset of $\mathbf{C}$ and let $u : \Omega \to \mathbf{R}$ be a continuous function. We say that u satisfies the *mean value property* if, for all $p \in \Omega$, the value $u(p)$ equals the average value of u on each circle centered at p and contained in Ω. Thus, for sufficiently small r,

$$(22) \qquad u(p) = \frac{1}{2\pi} \int_0^{2\pi} u(p + re^{i\theta}) d\theta.$$

Theorem 5.1. *A twice continuously differentiable function $u : \Omega \to \mathbf{R}$ is harmonic if and only if it satisfies the mean value property.*

Proof. Fix $p \in \Omega$, and consider the Taylor series expansion for u about p. We have, for small positive r,

$$(23) \qquad u(p + re^{i\theta}) = u(p) + \frac{\partial u}{\partial x}(p) r\cos(\theta) + \frac{\partial u}{\partial y}(p) r\sin(\theta)$$

$$+ \frac{1}{2}\frac{\partial^2 u}{\partial x^2}(p) r^2 \cos^2(\theta) + \frac{1}{2}\frac{\partial^2 u}{\partial y^2}(p) r^2 \sin^2(\theta)$$

$$+ \frac{\partial^2 u}{\partial x \partial y}(p) r^2 \cos(\theta)\sin(\theta) + E(r),$$

where $E(r)$ is an error term. To be more precise, we have $\lim_{r \to 0} \frac{E(r)}{r^2} = 0$. We compute the average value $A_p(u)$ of u on the circle by integrating, as in (22), the left-hand side of (23). Note that the integrals of $\cos^2$ and $\sin^2$ each equal π. The other integrals involving cosine and sine are zero. We obtain

(24) $$A_p(u) = \frac{1}{2\pi} \int_0^{2\pi} u(p + re^{i\theta})d\theta$$

$$= u(p) + \frac{r^2}{4}\left(\frac{\partial^2 u}{\partial x^2}(p) + \frac{\partial^2 u}{\partial y^2}(p)\right) + \frac{1}{2\pi}\int_0^{2\pi} E d\theta.$$

Since the error term tends to 0 faster than r^2, we see from (24) that $A_p(u) = u(p)$ if and only if

(25) $$\Delta(u)(p) = \left(\frac{\partial^2 u}{\partial x^2}(p) + \frac{\partial^2 u}{\partial y^2}(p)\right) = 0.$$

□

Our definition of harmonic function presumes that certain second derivatives exist. The mean value property provides an alternative equivalent definition that requires no differentiability assumptions. Harmonic functions turn out to be infinitely differentiable. We mention that harmonic functions in higher dimensions also arise in many scientific applications.

▶ **Exercise 5.19.** Prove Theorem 5.1 using partial derivatives with respect to z and $\overline{z}$.

▶ **Exercise 5.20.** Working with x, y, show that the real and imaginary parts of z, z^2, and z^3 are harmonic. Can you prove in this way that the real and imaginary parts of z^n are harmonic for all positive integers n? Compare with Exercise 5.6 and solve all these problems instantly!

▶ **Exercise 5.21.** Is the sum of harmonic functions harmonic? Under what conditions is the product of harmonic functions harmonic? If f is harmonic, under what condition is $\frac{\partial f}{\partial x}$ also harmonic?

▶ **Exercise 5.22.** Suppose that f is complex analytic on a connected open set and that $|f|^2$ is constant. Prove that f is constant. Do the same if $\mathrm{Re}(f)$ is constant. There are many proofs. Try to use only $\frac{\partial}{\partial z}$ and $\frac{\partial}{\partial \overline{z}}$.

▶ **Exercise 5.23.** By inspection find complex analytic functions whose real parts u are given by $u(x, y) = x^2 - y^2$, by $u(x, y) = e^x \cos(y)$, and by $u(x, y) = \frac{y}{x^2+y^2}$.

We close this section by providing a method for solving the natural problem from Exercise 5.23. Our method is considerably simpler than what appears in standard texts on complex analysis; a rigorous justification of our method relies on facts we will not prove. On the other hand, these facts are immediate for elementary functions, and hence our method is both computationally and conceptually simpler when u is given by a formula.

Given a harmonic function $u(x, y)$, defined in some region, we seek a complex analytic function f for which $u = \mathrm{Re}(f)$. As a consequence we also find a *conjugate harmonic function* $v(x, y)$ such that $f = u + iv$ is complex analytic.

Theorem 5.2. *Suppose $u = u(x, y)$ is harmonic in an open subset of $\mathbf{R}^2$ containing 0. Put $f(z) = 2u(\frac{z}{2}, \frac{z}{2i}) + c$ for an appropriate constant c. (In fact $c = -\overline{f(0)}$.) Then f is complex analytic and $u = \mathrm{Re}(f)$.*

Proof. By substituting the formulas for the real and imaginary parts, the equation $\mathrm{Re}(f(z)) = u(x, y)$ can be rewritten

$$(26) \qquad \frac{f(z) + \overline{f(z)}}{2} = \mathrm{Re}(f(z)) = u(x, y) = u\left(\frac{z + \overline{z}}{2}, \frac{z - \overline{z}}{2i}\right).$$

Formula (26) presumes that u is a real-analytic function; hence it is valid to replace x and y by their expressions in z and $\overline{z}$. These substitutions are valid, but a proof is somewhat sophisticated. See Theorem 7.1 of Chapter 8. For elementary functions such as polynomials, however, the substitutions are clearly valid. Now comes the key point. Formula (26) is an identity that holds for all z and $\overline{z}$. Hence we can treat z and $\overline{z}$ as independent variables. In (26) we substitute $\overline{z} = 0$, while keeping z free. We obtain

$$(27) \qquad \frac{f(z) + \overline{f(0)}}{2} = u(x, y) = u\left(\frac{z}{2}, \frac{z}{2i}\right).$$

The theorem follows immediately from (27). $\qquad\qquad\qquad\qquad\qquad\qquad\square$

In case 0 is not in the domain of u, we can still use the same method by substituting another value for $\overline{z}$.

Many texts on complex variables provide a different method for solving this problem; the method involves both differentiating and integrating, using the Cauchy-Riemann equations. Here is that method. Find the partial derivative u_x and set it equal to v_y. Integrate to get $v(x, y) = \int v_y(x, y)dy + \phi(x)$ for some integration constant ϕ. Then use $v_x = -u_y$ to determine $\phi'(x)$ and hence to determine ϕ. Finally put $f = u + iv$ and simplify.

The point of the next several exercises is to compare the efficiency of the methods. In two of these exercises you cannot set $\overline{z}$ equal to 0; choose another point.

▶ **Exercise 5.24.** Given $u(x, y) = x^2 - y^2$, find a complex analytic f such that $u = \mathrm{Re}(f)$ by using Theorem 5.2. Then find f by the standard method.

▶ **Exercise 5.25.** Do the same problem for $u(x, y) = x^3 - 3xy^2$.

▶ **Exercise 5.26.** Do the same problem for $u(x, y) = e^x \cos(y)$.

▶ **Exercise 5.27.** Do the same problem for $u(x, y) = \log(x^2 + y^2)$. Assume $x, y > 0$.

▶ **Exercise 5.28.** Do the same problem for $u(x, y) = \frac{y}{x^2 + y^2}$.

The next few exercises require a bit more effort. The purpose of the first one is to look ahead to the Cauchy integral formula to establish the mean value property for the real and imaginary parts of a complex analytic function.

▶ **Exercise 5.29** (Mean value property). Suppose that f is complex analytic on **C** and $f = u + iv$ as usual. For each z and each positive radius, use the Cauchy integral formula from Chapter 6 to establish the mean value property (22) for u and for v.

▶ **Exercise 5.30.** Show, using the mean value property, that a harmonic function cannot achieve its maximum (or minimum) in an open set unless it is a constant. This property is called the maximum principle and has many applications in pure and applied mathematics.

▶ **Exercise 5.31.** If $\Delta(u) = 0$, show that the real Hessian matrix of second derivatives of u must have a nonpositive determinant.

The next two exercises show that $|f|$ satisfies the maximum principle when f is complex analytic.

▶ **Exercise 5.32.** A smooth function u is called *subharmonic* if $\Delta(u) \geq 0$. Show that $u(p) \leq A_p(u)$ for all p and all circles about p if and only if $\Delta(u) \geq 0$. Then verify the maximum principle for subharmonic functions.

▶ **Exercise 5.33.** Suppose that f is complex analytic. Show that $|f|$ and $|f|^2$ are subharmonic. Harder: Show, assuming that $f \neq 0$, that $\log(|f|^2)$ is subharmonic. See pages 198–207 of [**7**] for related ideas.

▶ **Exercise 5.34.** Find all functions f such that f is complex analytic on $\mathbf{C}$ and that $f(|z|) = |f(z)|$ for all z.

The Dirichlet problem. Several of the most profound applications of complex analysis to the sciences follow from the relationship between complex analytic functions and harmonic functions. In particular complex analysis provides an approach to the Dirichlet problem, which is one of the most important boundary value problems in pure and applied mathematics. We discuss these matters briefly in this section and refer to [**1, 10, 18, 23**] for more information about both the mathematics and its applications.

Let Ω be an open connected set in $\mathbf{C}$ and suppose that its boundary $b\Omega$ is a reasonably smooth object. Let $g : b\Omega \to \mathbf{R}$ be a continuous function. The *Dirichlet problem* for Ω is to find a function u such that u is harmonic on Ω and $u = g$ on $b\Omega$. We give a simple example. Let Ω be the unit disk and let $g(\theta) = \cos(\theta)$. Put $u(x, y) = x = \mathrm{Re}(z)$. Thus u is harmonic. On the boundary circle we have $z = e^{i\theta} = \cos(\theta) + i\sin(\theta)$ and hence $u = g$ there.

The same method solves the Dirichlet problem for the unit disk when g is $\cos(n\theta)$ or $\sin(n\theta)$ for any integer n, as we next indicate. More generally, for $0 \leq j \leq N$, let c_j be complex numbers and put $c_{-j} = \overline{c_j}$. In particular c_0 is real. The function g in (28) is then real-valued and continuous. We can solve the Dirichlet problem for the unit disk whenever g has the form

$$(28) \qquad\qquad g(\theta) = \sum_{k=-N}^{N} c_k e^{ik\theta}.$$

We define $u(x, y)$ by

$$(29) \qquad u(x, y) = c_0 + \sum_{k=1}^{N} 2\mathrm{Re}(c_k z^k) = 2\mathrm{Re}\left(\frac{c_0}{2} + \sum_{k=1}^{N} c_k z^k\right).$$

Since u is the real part of a polynomial in z, u is harmonic. On the other hand, suppose $|z| = 1$. Then $z = e^{i\theta}$ and we obtain

$$(30) \qquad u(x, y) = c_0 + \sum_{k=1}^{N} 2\mathrm{Re}(c_k e^{ik\theta}) = c_0 + \sum_{k=1}^{N} c_k e^{ik\theta} + \sum_{k=1}^{N} \overline{c_k} e^{-ik\theta}.$$

Since $\overline{c_k} = c_{-k}$, we may replace k by $-k$ in the second sum and obtain (28). Thus $u = g$ on the boundary, and hence u solves the Dirichlet problem.

The method of the previous paragraph also works by replacing g in (28) with a convergent infinite sum of the same kind. This infinite sum is called the Fourier series of g. When g is continuous, it has a Fourier series.

It is also possible to give an explicit integral formula, called the Poisson integral formula, which solves the Dirichlet problem in the unit disk. We state the formula and refer to any of [1, 10, 18, 23] for derivations. Given a continuous function g on the unit circle, define u by the following integral. Then u is harmonic in the unit disk and $u = g$ on the circle:

$$(31) \qquad u(w) = \frac{1}{2\pi} \int_{|z|=1} \frac{1 - |w|^2}{|z - w|^2} g(z) d\theta.$$

At first glance it might seem too special to solve the Dirichlet problem in the unit disk. It turns out, however, that we can then solve it for any simply connected domain with nice boundary. The details are beyond the scope of this book, but we sketch some of the ideas here.

Let Ω be a connected open subset of $\mathbf{C}$ and let $\mathbf{D}$ denote the unit disk. Suppose we can find a complex analytic function $F : \mathbf{D} \to \Omega$ such that F is a bijection and F^{-1} is also complex analytic. Such a function F is called a conformal map from $\mathbf{D}$ to Ω. Such mappings F exist in considerable generality; see the Riemann mapping theorem (Theorem 5.1 of Chapter 8). In favorable (but still quite general) circumstances, F extends to be continuous on the unit circle and defines a bijection of the circle to $b\Omega$. We can then solve the Dirichlet problem for Ω. Given g, we put $G = g \circ F$. Then G is continuous on the unit circle. We find a harmonic U on the unit disk with $U = G$ on the circle. Put $u = U \circ F^{-1}$. We claim that u solves the Dirichlet problem on Ω. On the boundary

$$u = G \circ F^{-1} = (g \circ F) \circ F^{-1} = g.$$

On the other hand, we claim that u is harmonic. Put $H = F^{-1}$ to simplify notation. The claim follows from the following simple yet often applied result.

Lemma 5.1. *Let Ω_1 and Ω_2 be open subsets of $\mathbf{C}$. Assume $H : \Omega_2 \to \Omega_1$ is complex analytic and U is harmonic on Ω_1. Then $u = U \circ H$ is harmonic on Ω_2.*

Proof. Let z denote the variable in Ω_1 and let w denote the variable in Ω_2. Using (21), we compute the Laplacian $4u_{w\overline{w}}$ where the subscripts denote partial derivatives. Note that $H_{\overline{w}} = 0$ by the Cauchy-Riemann equations. We are given that $U_{z\overline{z}} = 0$. By the chain rule we therefore get

$$u_{w\overline{w}} = (U_z H_w)_{\overline{w}} = U_{z\overline{z}} |H_w|^2 = 0.$$

$\square$

The importance of the Dirichlet problem in applications has led to many different methods in analysis. See [1] for a solution using subharmonic functions.

▶ **Exercise 5.35.** Suppose h is complex analytic and $h : \mathbf{R} \to \mathbf{R}$. Assume u is harmonic. True or false? $h \circ u$ is harmonic. How does this situation differ from Lemma 5.1?

▶ **Exercise 5.36.** Using a linear fractional transformation that maps **D** bijectively to the upper half-plane, apply the discussion in this section to solve the Dirichlet problem in the upper half-plane.

▶ **Exercise 5.37.** Verify the following identity, which arises in proving the Poisson integral formula:

$$\frac{z}{z-w} + \frac{\overline{w}}{\overline{z}-\overline{w}} = \frac{|z|^2 - |w|^2}{|z-w|^2}.$$

▶ **Exercise 5.38.** (Difficult) Fill in the details of the following derivation of the Poisson integral formula. Given g, the value of $u(0)$ is determined by the mean value property (22). Next, given w in the unit disk, find a linear fractional transformation L that maps the circle to the circle and maps w to 0. A formula for such an L appears in Exercise 2.4 and in formula (28) from Chapter 8. Given L, use the previous discussion and Lemma 5.1 to determine $u(w)$.

6. What is a differential form?

Students first see dx and dy in beginning calculus, but the precise meaning of these symbols is never explained. Some students have been taught to say that dx is an *infinitesimal change*; most such students get all confused when asked to elaborate. Precise mathematics should be expressed using sets, functions, and logical symbols. What kind of object is dx?

In a calculus course it feels misguided to define differential forms precisely. Excessive pedantry at that time of one's educational development serves to hinder progress. Let us therefore introduce differential forms in a rigorous but informal manner.

We work in Euclidean space $\mathbf{R}^n$; restricting to $n = 1$ and $n = 2$ makes it harder rather than easier to understand. Let Ω be an open set in $\mathbf{R}^n$, and let $p \in \Omega$. A tangent vector $\mathbf{v}_p$ at p is simply an element of $\mathbf{R}^n$ associated with p. When p is understood, we write $\mathbf{v}$ instead of $\mathbf{v}_p$. Given such a $\mathbf{v}$, we use it to find directional derivatives. Thus, if f is a differentiable function defined near p, we compute the directional derivative

$$\frac{\partial f}{\partial \mathbf{v}}(p) = \lim_{t \to 0} \frac{f(p + t\mathbf{v}) - f(p)}{t}.$$

This number represents the rate of change of f if we move from p in the direction $\mathbf{v}$. Note that $p + t\mathbf{v}$ gives a parametric equation for the line through p with direction $\mathbf{v}$. We are doing one-variable calculus along this line.

We define the differential df by the rule

(32)
$$df(p)(\mathbf{v}) = \frac{\partial f}{\partial \mathbf{v}}(p).$$

Notice the similarity between $df(p)$ and the gradient vector $\nabla f(p)$. We can express (32) using the gradient vector or in coordinates as

(33)
$$df(p)(\mathbf{v}) = \nabla f(p) \cdot \mathbf{v} = \sum_{j=1}^{n} \frac{\partial f}{\partial x_j}(p) v_j.$$

Linearity follows from (33):

$$df(p)(\mathbf{v} + \mathbf{w}) = df(p)(\mathbf{v}) + df(p)(\mathbf{w}),$$

$$df(p)(c\mathbf{v}) = c \, df(p)(\mathbf{v}).$$

Thus df is a machine that assigns to each tangent vector $\mathbf{v}$ at p the number defined in (32). Hence df is a function; its domain consists of pairs $(p, \mathbf{v}_p)$. The value of df at this pair is written $df(p)(\mathbf{v}_p)$. Often we think of p as fixed. Then $df(p)$ is a linear function that assigns a number to each $\mathbf{v}_p$. In the language of linear algebra, $df(p)$ is an element of the *dual space* to the space T_p of tangent vectors at p. We can add such differentials and multiply them by scalars. The more surprising thing is that there is an interesting way to multiply them by each other, called the *exterior product* or the *wedge product*. The wedge product sheds light on the theory of determinants.

Let V be any finite-dimensional vector space over the real or complex numbers. Given linearly independent vectors $e_1,...,e_k$ in V, we define a new object, called a k-form, written in (34):

$$(34) \qquad\qquad e_1 \wedge e_2 \wedge ... \wedge e_k.$$

We decree that the object in (34) gets multiplied by -1 if we switch two of the factors. Hence, for a permutation $j_1,...,j_k$, we have

$$e_{j_1} \wedge e_{j_2} \wedge ... \wedge e_{j_k} = \pm e_1 \wedge e_2 \wedge ... \wedge e_k,$$

where the sign equals the *signum* of the permutation.

Suppose the e_j form a basis for V. Given vectors $v_i \in V$, for $1 \le i \le k$, then there are unique scalars v_{ij} such that we can write

$$v_i = \sum_j v_{ij} e_j.$$

We define the wedge product by

$$(35) \qquad\qquad v_1 \wedge v_2 \wedge ... \wedge v_k = \det(v_{ij}) e_1 \wedge e_2 \wedge ... \wedge e_k.$$

We recall the basic facts about determinants. In sophisticated language, the determinant is an *alternating multilinear form*. Think of the determinant as a function of its rows. It is linear in each row if the other rows are fixed, and it changes sign if two rows are switched. The determinant function is uniquely determined by these properties together with the normalization that the determinant of the identity matrix is 1.

Hence the wedge product is linear in each v_i if the other vectors are fixed and it switches signs if we switch two of the vectors. The wedge product of k vectors in a k-dimensional space is 0 if and only if the vectors are linearly dependent.

The crucial point is the geometric interpretation of determinants in terms of volume. If we assume that the basis vectors $e_1,...,e_k$ span the unit box, then the oriented volume of the box spanned by $v_1,...,v_k$ equals $\det(v_{ij})$. This approach allows us to regard $e_1 \wedge e_2 \wedge ... \wedge e_k$ as a *volume form* on k-dimensional space.

▶ **Exercise 5.39.** Show by elementary math that the oriented area of the parallelogram spanned by $a + bi$ and $c + di$ in $\mathbf{C}$ is the determinant $ad - bc$.

▶ **Exercise 5.40.** Let $\omega = Pdx + Qdy$. Define $*\omega$ by $*\omega = -Qdx + Pdy$. This operation gives a special case of the Hodge $*$ operator. If $\omega = du$, where u is harmonic, what is $*\omega$? What does this exercise have to do with orthogonal trajectories?

▶ **Exercise 5.41.** Define the Hodge $*$ operator more generally by $*1 = dx \wedge dy$ and $*(dx \wedge dy) = 1$, and extend by linearity. Thus for example $*g = g \, dx \wedge dy$. Let f be a smooth function. What is $*d * df$?

▶ **Exercise 5.42.** Let $\omega = Pdx + Qdy$ be a 1-form, but regard it as a vector field $L = (P, Q)$. Express the divergence and curl of this vector field in terms of d and $*$.

▶ **Exercise 5.43.** In real four-dimensional space put $\eta = dx_1 \wedge dx_2 + dx_3 \wedge dx_4$. Find $\eta \wedge \eta$. In complex two-dimensional space, put $\omega = dz_1 \wedge d\bar{z}_1 + dz_2 \wedge d\bar{z}_2$. Find $\omega \wedge \omega$.

Chapter 6

Complex Integration

Integrals play a crucial role in calculus, engineering, and physics. It is therefore not unexpected that integrals will be crucial to our development of complex geometry. We will use complex line integrals to prove one of the most fundamental facts in complex analysis: a complex analytic function is given locally by a convergent power series. In doing so, we will establish that the three possible definitions of complex analytic function given in Chapter 5 all yield the same class of functions.

Complex integration has many other applications. In particular we show how to find various real definite integrals by using complex integration. Doing so allows us in Chapter 7 to glimpse such topics as Fourier transforms, the Gaussian probability distribution, and the Γ-function.

1. Complex-valued functions

We first discuss some basic facts about derivatives and integrals of complex-valued functions of one real variable. Most readers will blithely accept and use these facts without much thought.

Let (a, b) be an open interval on $\mathbf{R}$ and suppose that $g : (a, b) \to \mathbf{C}$ is a function. Write $g(t) = u(t) + iv(t)$ for its expression in terms of real and imaginary parts. Then g is differentiable if and only if both u and v are differentiable, and we have $g'(t) = u'(t) + iv'(t)$ for all $t \in (a, b)$. The derivative is linear over $\mathbf{C}$. In other words, if $c \in \mathbf{C}$ and g and h are differentiable complex-valued functions on (a, b), then cg and $g + h$ also are, and of course we have

$$(cg)'(t) = cg'(t),$$

$$(g + h)'(t) = g'(t) + h'(t).$$

Next let $[a, b]$ be a closed interval on $\mathbf{R}$, and let $g : [a, b] \to \mathbf{C}$ be a function. As above, write $g(t) = u(t) + iv(t)$ for its expression in terms of real and imaginary parts. Note that g is continuous if and only if u and v are continuous. When g is continuous, it is natural to call its image a curve. Following common practice, we use the term *curve* both for the mapping g and for the image $g([a,b])$ as a subset of $\mathbf{C}$. We reserve the term curve for continuous mappings g. A curve is called *smooth* (or *continuously differentiable*) if g is smooth (or continuously differentiable). A curve is called *piecewise smooth* if it has finitely many smooth pieces. A piecewise smooth curve is connected when regarded as a subset of $\mathbf{C}$.

The definition of the integral of a continuous (or, more generally, piecewise continuous) function g reduces to the usual definition of a real-valued function as follows:

$$(1) \qquad \int_a^b g = \int_a^b g(t)dt = \int_a^b u(t)dt + i \int_a^b v(t)dt = \int_a^b u + i \int_a^b v.$$

Said another way, the definition states

$$(2) \qquad \operatorname{Re} \int g = \int \operatorname{Re}(g),$$

$$(3) \qquad \operatorname{Im} \int g = \int \operatorname{Im}(g).$$

The definitions (2) and (3) are reasonable. The real part of a sum is the sum of the real parts of the summands. The real part of a limit is the limit of the real part. The same holds for the imaginary part. An integral is the limit of a sum. Therefore we can move the operation of taking the real or imaginary part past the integral sign.

It follows from (1) that the integral is linear over $\mathbf{C}$. In other words, if $c \in \mathbf{C}$ and g and h are continuous complex-valued functions on $[a,b]$, then

$$\int_a^b cg = \int_a^b cg(t)dt = c \int_a^b g(t)dt = c \int_a^b g,$$

$$\int_a^b (g+h) = \int_a^b (g(t) + h(t))dt = \int_a^b g(t)dt + \int_a^b h(t)dt = \int_a^b g + \int_a^b h.$$

These formal basic facts about the linearity of derivatives and integrals hold as we expect. We say a bit more about a crucial inequality.

First of all, integrals (in any theory of integration) are limits of sums and therefore integrals of real-valued functions preserve inequalities. If $u : [a,b] \to \mathbf{R}$ is an integrable function and $0 \le u$, then $0 \le \int_a^b u$. It follows by linearity that if u_1 and u_2 are integrable functions with $u_1 \le u_2$, then $\int_a^b u_1 \le \int_a^b u_2$. Since both $u \le |u|$ and $-u \le |u|$ always hold, we obtain both $\pm \int u \le \int |u|$. Hence for each real-valued integrable function u we have

$$(4) \qquad \left| \int_a^b u \right| \le \int_a^b |u|.$$

This inequality also holds for complex-valued functions, but requires justification.

Lemma 1.1. *Inequality (4) holds for each integrable complex-valued function g.*

Proof. We write $\int_a^b g(t)dt = re^{i\theta}$. We then have

$$(5) \quad \left| \int_a^b g(t)dt \right| = r = e^{-i\theta} \int_a^b g(t)dt = \int_a^b e^{-i\theta} g(t)dt = \text{Re} \int_a^b e^{-i\theta} g(t)dt$$

$$= \int_a^b \text{Re}\left(e^{-i\theta} g(t)\right) dt \le \int_a^b |e^{-i\theta} g(t)|dt = \int_a^b |g(t)|dt.$$

In one step in (5) we replaced an integral by its real part; that step is fine, because $|\int g|$ is real. Also, with $h = e^{-i\theta}g$, we integrated the inequality $\text{Re}(h) \le |h|$ to obtain $\int_a^b \text{Re}(h) \le \int_a^b |h|$. In the last step we used $|e^{-i\theta}| = 1$. $\square$

2. Line integrals

We begin with the notion of work in elementary physics. Imagine a force acting at each point of the plane. We can think of this force as an arrow based at each point; in other words, this force defines a vector field. A vector field is a function F, in general not complex analytic, mapping $\mathbf{C}$ to $\mathbf{C}$. We write $F = P + iQ = (P, Q)$ in terms of its real components. At each point $z \in \mathbf{C}$, we place the vector $F(z)$.

Given a curve γ in the plane satisfying appropriate hypotheses, we ask how much work it takes to travel along this curve against the force F. Work is the dot product of force and distance. We approximate the curve γ by a piecewise rectangular path whose sides are parallel to the axes. We resolve the force into components and compute the work along these sides by the simpler formula of force times distance. Finally we add together the infinitesimal pieces, resulting in a line integral (or path integral). Hence work should be the integral along γ of $Pdx + Qdy$. An alternative notation used in physics helps clarify the picture. By definition the work W is defined by the line integral in (6):

$$(6) \qquad\qquad W = \int_\gamma F \cdot d\mathbf{s}.$$

We can think of $d\mathbf{s}$ as the vector (dx, dy). We write ds for the length of $d\mathbf{s}$. Before giving a precise definition of the meaning of (6), we recall from calculus the notion of arc-length. The length $L = L(\gamma)$ of a smooth curve γ is defined by

$$(7) \qquad\qquad L = L(\gamma) = \int_\gamma ds = \int_a^b |\gamma'(t)|dt.$$

This definition arises from approximating the curve by a rectangular path as above and using the Pythagorean theorem. We say that the curve γ is *smooth* if the function $t \to \gamma(t)$ is smooth. When γ is smooth, the far right-hand side of (7) has a precise meaning. We then *define* $\int_\gamma ds$ by (7). When γ is piecewise smooth, we use (7) on each smooth piece and then add up the results. In this book we will nearly always assume that our curves are piecewise smooth.

In the notation from physics, ds is the object which becomes $|\gamma'(t)|dt$ after we parametrize γ. Its vector form $d\mathbf{s} = (dx, dy)$ from (6) becomes $\gamma'(t)dt$ after we parametrize γ. If $F = (P, Q)$, then (6) becomes

$$(8) \qquad W = \int_\gamma F \cdot d\mathbf{s} = \int_\gamma P dx + Q dy.$$

Mathematicians prefer expressing these ideas using differential forms. Let $\gamma :$ $[a, b] \to \mathbf{R}^2$ be a smooth curve, and let $\omega = P dx + Q dy$ be a smooth 1-form defined near the image of γ. We write $\gamma(t) = (x(t), y(t))$ and we define the integral by (9):

$$(9) \qquad \int_\gamma \omega = \int_\gamma P dx + Q dy = \int_a^b \left(P(\gamma(t)) x'(t) + Q(\gamma(t)) y'(t) \right) \, dt.$$

Again, when γ is piecewise smooth, we use (9) on each piece and add up the results. One technical point must be established for these definitions to make sense. We must show that the value of the line integral $\int_\gamma P dx + Q dy$ is independent of the parametrization of γ.

Lemma 2.1. *Formula* (9) *is unchanged if we reparametrize* γ. *In other words, suppose* $\gamma : [a, b] \to \mathbf{C}$ *is a smooth curve and* $\phi : [a', b'] \to [a, b]$ *is a continuously differentiable bijection. Put* $\eta = \gamma \circ \phi$. *Then*

$$(10) \qquad \int_\gamma P dx + Q dy = \int_\eta P dx + Q dy.$$

Proof. The proof is left to the reader, with the following hint. Write the left-hand side of (10) using (9). Then put $t = \phi(\tau)$. Use the change of variables formula from elementary calculus to obtain the right-hand side of (10). $\qquad\square$

We therefore consider 1-forms to be the basic objects. Formula (9) tells us how to compute a line integral, and formula (8) interprets the integral in terms of the work done.

▶ **Exercise 6.1.** Fill in the details of the proof of Lemma 2.1.

The following inequality about length generalizes Lemma 1.1 and gets used in some proofs.

Proposition 2.1 (ML-inequality). *Let* $\gamma : [a, b] \to \mathbf{R}^2$ *be a piecewise smooth curve of length* L. *Assume that* f *is a function, defined and continuous on the image of* γ, *with* $|f(z)| \le M$. *Then*

$$\left| \int_\gamma f(z) dz \right| \le ML.$$

Proof. It suffices to prove the inequality for each smooth piece of γ and to add the results. We therefore assume γ is smooth. The definition of the line integral and Lemma 1.1 give

$$\left| \int_\gamma f(z) dz \right| = \left| \int_a^b f(\gamma(t)) \gamma'(t) dt \right| \le \int_a^b |f(\gamma(t))| |\gamma'(t)| dt \le M \int_a^b |\gamma'(t)| dt = ML.$$

$\qquad\square$

We next discuss Green's theorem. We prove the theorem only when γ is the boundary of a rectangle and refer to any multi-variable calculus book for a proof in more generality. We mention however that the proof for a rectangle is in some sense the main point. Figures 6.2 and 6.3 suggest that the result for rectangles implies the result for any region bounded by finitely many straight lines, each parallel to one of the axes. As indicated by the arrows in Figure 6.2, a line integral over the boundary of the large rectangle equals the sum of line integrals over the boundaries of the four smaller rectangles, because the contributions over the interior segments cancel in pairs. Also, a double integral over the large rectangle equals the sum of the double integrals over the four pieces. It follows, as suggested by Figure 6.3, that the conclusion of Green's theorem then also holds for any region bounded by a polygonal path whose sides are parallel to the axes. Finally, again suggested by Figure 6.3, the theorem holds for any curve that is a limit of such polygonal paths. This collection of limit curves includes every curve that arises in this book; in particular, piecewise smooth curves are such limits.

Recall that a closed curve is a function $\gamma : [a, b] \to \mathbf{C}$ with $\gamma(a) = \gamma(b)$. We say that a closed curve γ is *simple* if the image does not cross itself, except at the end points. In other words, if $s \neq t$, then $\gamma(s) \neq \gamma(t)$ unless $s = a$ and $t = b$ or $s = b$ and $t = a$. We say that a closed curve γ is *positively oriented* if the image is traversed counterclockwise as the parameter t increases from a to b. See Figure 6.1. (A more rigorous definition is possible, but we are content with this intuitive definition.)

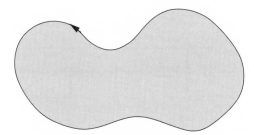

Figure 6.1. A positively oriented simple closed curve enclosing a region.

Theorem 2.1 (Green's theorem). *Let γ be a piecewise smooth, positively oriented, simple closed curve in $\mathbf{C}$. Let P, Q be continuously differentiable functions on and inside γ. Write Ω for the interior of γ. Then*

$$(11) \qquad \int_\gamma P dx + Q dy = \iint_\Omega \left(\frac{\partial Q}{\partial x} - \frac{\partial P}{\partial y} \right) dx dy.$$

Proof. We assume γ is the boundary of a rectangle Ω, with vertices at $(a, b), (c, b),$ $(c, \delta), (a, \delta)$. Since γ is positively oriented, the x variable increases from a to c when $y = b$, and decreases from c to a when $y = \delta$. Similarly the y variable increases from b to δ when $x = c$ and decreases from δ to b when $x = a$.

We integrate the right-hand side of (11) over the interior Ω of this rectangle and use the fundamental theorem of calculus. We may do the iterated integral in either order (Fubini's theorem); we integrate $\frac{\partial Q}{\partial x}$ first with respect to x, and we integrate $\frac{\partial P}{\partial y}$ first with respect to y. We obtain

$$(12) \quad \iint_\Omega \left(\frac{\partial Q}{\partial x} - \frac{\partial P}{\partial y} \right) dxdy = \int_b^\delta \int_a^c \frac{\partial Q}{\partial x} dxdy - \int_a^c \int_b^\delta \frac{\partial P}{\partial y} dydx$$

$$= \int_b^\delta (Q(c,y) - Q(a,y))dy - \int_a^c (P(x,\delta) - P(x,b))dx = \int_\gamma Qdy + Pdx.$$

The last step follows simply by parametrizing the given line segments and keeping track of the orientation, as in the first paragraph. $\qquad\square$

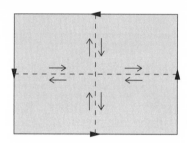

Figure 6.2. Breaking a rectangle into smaller rectangles.

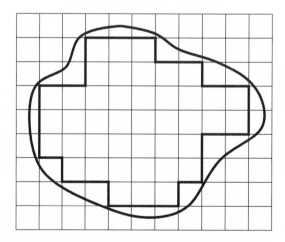

Figure 6.3. Approximation of a curve by a polygonal path.

Green's theorem is a special case of what has become known as the modern form of Stokes's theorem. Although this material is a bit too advanced for this book, at the end of this section we briefly whet the reader's appetite for learning about differential forms. The simplest formulation of Green's theorem uses the operator d, called the exterior derivative. We summarize the properties of d at the end of this section. We next restate Green's theorem using the exterior derivative.

Each 1-form in the plane can be written $Pdx + Qdy$. We use the properties of the exterior derivative from Proposition 2.2 at the end of this section. Assuming P and Q are differentiable, we have

$$d(Pdx + Qdy) = \left(\frac{\partial P}{\partial x}dx + \frac{\partial P}{\partial y}dy\right) \wedge dx + \left(\frac{\partial Q}{\partial x}dx + \frac{\partial Q}{\partial y}dy\right) \wedge dy$$

$$= \frac{\partial P}{\partial y}dy \wedge dx + \frac{\partial Q}{\partial x}dx \wedge dy = \left(-\frac{\partial P}{\partial y} + \frac{\partial Q}{\partial x}\right) dx \wedge dy.$$

The formula in Green's theorem becomes

$$(13) \qquad \int_\gamma Pdx + Qdy = \iint_\Omega d(Pdx + Qdy).$$

Next we investigate the use of Green's theorem in complex analysis.

Corollary 2.1 (Easy form of Cauchy's theorem). *Let γ and Ω be as in Green's theorem, and suppose f is complex analytic on a set containing the closure of Ω. Then*

$$\int_\gamma f(z)dz = 0.$$

Proof. Write $f = u + iv$ and $dz = dx + idy$. Then, using Green's theorem and the Cauchy-Riemann equations, we get

$$\int_\gamma f(z)dz = \int_\gamma (u + iv)(dx + idy) = \int_\gamma (udx - vdy) + i\int_\gamma (vdx + udy)$$

$$= \iint_\Omega (\frac{-\partial v}{\partial x} - \frac{\partial u}{\partial y})dxdy + i\iint_\Omega (\frac{\partial u}{\partial x} - \frac{\partial v}{\partial y})dxdy = 0 + 0 = 0.$$

$\square$

Students are sometimes careless in applying these results. We give two (related) examples where the differentiability hypotheses fail only at a single point in the interior of the curve, and the conclusion fails. These two examples lie at the basis of the mathematical subject called *cohomology*. Cohomology makes precise the number of and the dimension of holes in a complicated object. Similar examples in three dimensions lie at the basis of the theory of electricity and magnetism.

Example 2.1. Put $\omega = \frac{-y}{x^2+y^2}dx + \frac{x}{x^2+y^2}dy$ for $(x,y) \neq (0,0)$. Let γ be the positively oriented unit circle. Then $\int_\gamma \omega = 2\pi$. On the other hand, for $(x,y) \neq (0,0)$, we have $d\omega = 0$. These statements together do not contradict Green's theorem. Its hypotheses do not apply because ω is not differentiable at 0.

Example 2.2. Suppose $f(z) = \frac{1}{z}$ and γ is the unit circle. Then Cauchy's theorem does not apply. In fact, $\int_\gamma f(z)dz = 2\pi i \neq 0$. The reason is that the hypotheses in Green's theorem insist that something is true everywhere in the interior Ω of γ. If $f(z) = \frac{1}{z}$, then f fails to be analytic at 0. In other words, one must be careful. Simply being complex analytic *near* γ is not good enough!

▶ **Exercise 6.2.** With ω as in Example 2.1, prove that $d\omega = 0$. Then express ω in terms of dz and $d\bar{z}$ and prove it again.

▶ **Exercise 6.3.** Evaluate the line integrals in the two previous examples.

We have often mentioned that things are easier if we work with z and $\bar{z}$ rather than with x and y. That comment applies particularly well here. We write out the proof of Corollary 2.1 using complex derivatives:

$$(14) \qquad \int_\gamma f(z)dz = \iint_\Omega d(f(z)dz) = \iint_\Omega \frac{\partial f}{\partial \bar{z}}(z)d\bar{z} \wedge dz = \iint_\Omega 0 = 0.$$

These ideas are closely related to physics. Consider a vector field $F = (P, Q)$ in the plane. This vector field F is called *conservative* if the work done traveling along any path depends on only the difference in potential energy. In other words, the line integral $\int_\gamma F \cdot d\mathbf{s}$ depends on only the endpoints of γ. Mathematicians express the same idea using differential forms. Let $\omega = Pdx + Qdy$ be a 1-form. We say that ω is an *exact* 1-form if there is a smooth function g such that $dg = Pdx + Qdy$. For any piecewise smooth curve γ, whether closed or not, we then have

$$\int_\gamma Pdx + Qdy = \int_\gamma dg = g(\gamma(b)) - g(\gamma(a)).$$

Thus, if ω is exact, then F is conservative.

The following standard result summarizes the situation. In it we use a topological fact about connected open sets in $\mathbf{C}$; its proof is sketched in Section 6 of Chapter 1. We can connect any two points in Ω by a polygonal path whose sides are parallel to the axes.

Theorem 2.2. *Let Ω be a connected open set on which $\omega = Pdx + Qdy$ is smooth. The line integral $\int_\gamma \omega$ depends on only the endpoints of γ if and only if there is a smooth function g on Ω for which $\omega = dg$.*

Proof. If g exists, then as above, $\int_\gamma dg = g(\gamma(b)) - g(\gamma(a))$. To prove the converse, fix a point z_0 in Ω. Given $z \in \Omega$, we connect z_0 to z by a polygonal path γ whose sides are parallel to the axes. We define g by $g(z) = \int_\gamma \omega$. By assumption the value of g is independent of the choice of path. By choosing the last part of the polygonal path to be parallel to the x-axis and using the fundamental theorem of calculus, we see that $\frac{\partial g}{\partial x} = P$. By choosing the last part of the polygonal path to be parallel to the y-axis, we likewise see that $\frac{\partial g}{\partial y} = Q$. $\qquad\square$

When $f(z) = z^{n-1}$ for $n \in \mathbf{N}$, we have $f(z)dz = d(\frac{z^n}{n})$, and thus $f(z)dz$ is exact. When γ is a closed curve, we therefore obtain

$$(15) \qquad \int_\gamma z^{n-1}dz = \int_\gamma d(\frac{z^n}{n}) = 0.$$

By linearity and (15), for any polynomial p we obtain

$$\int_\gamma p(z)dz = 0.$$

It is therefore no surprise that the same result for convergent power series holds. Let F be complex analytic; then $dF = F'(z)dz$ is an exact differential. Therefore $\int_\gamma F'(z)dz = 0$ for any closed curve γ. We obtain Cauchy's theorem for derivatives of complex analytic functions.

▶ **Exercise 6.4.** In each case find parametric equations $t \to z(t)$ for the curve γ. Use complex variable notation rather than real variable notation in 1) and 2).

1) γ is the unit circle, traversed counterclockwise.

2) γ is a circle of radius R with center at p, traversed counterclockwise.

3) γ is the boundary of a rectangle with vertices at $(a, b), (c, b), (c, d), (a, d)$, traversed counterclockwise. Four different formulas are required.

4) γ is the part of the right-hand branch of the hyperbola defined by $x^2 - y^2 = 3$ from $(2, -1)$ to $(2, 1)$.

5) γ is the ellipse $\frac{x^2}{a^2} + \frac{y^2}{b^2} = 1$, traversed counterclockwise.

▶ **Exercise 6.5.** Graph the set of points in the plane where $x^3 + y^3 = 3xy$. This set is called the folium of Descartes. Let γ be the part of the curve forming a loop in the first quadrant. 1) Find parametric equations for γ. Hint: Determine where the line $y = tx$ intersects γ to find x and y as functions of t.

2) Find the area inside the loop. See Figure 6.4. Hint: First express $\frac{x\,dy - y\,dx}{2}$ in terms of t and then integrate it around γ.

▶ **Exercise 6.6.** For $k \in \mathbf{N}$, define a curve by $x^{2k+1} + y^{2k+1} = (2k + 1)x^k y^k$. Repeat Exercise 6.5 for these curves.

▶ **Exercise 6.7.** Let $p_d(x, y)$ be a homogeneous polynomial of degree d, and let $q_{d-1}(x, y)$ be homogeneous of degree $d - 1$. Consider the set V of (x, y) such that $p_d(x, y) = q_{d-1}(x, y)$. Parametrize V by proceeding as in the previous exercises.

▶ **Exercise 6.8.** Evaluate, for $n \in \mathbf{Z}$, the integral $\int_{|z|=R} z^n d\bar{z}$.

▶ **Exercise 6.9.** Use polar coordinates to evaluate, for nonnegative integers m, n, the double integral

$$\int_{|z| \leq R} z^m \bar{z}^n dz \, d\bar{z}.$$

Exterior derivative. As we saw above, Green's theorem has a simple statement using the exterior derivative d. We glimpse deeper waters by summarizing this concept in n dimensions. We mention first that a smooth function can be regarded as a differential 0-form. For each k-form ϕ we define a $(k + 1)$-form $d\phi$ called the *exterior derivative* of ϕ. The operator d is defined via the following result:

Proposition 2.2. *There is a unique function d mapping k-forms to $(k + 1)$-forms and satisfying the following properties:*

- $d(\phi + \psi) = d\phi + d\psi$.

- $d(\phi \wedge \psi) = d\phi \wedge \psi + (-1)^p \phi \wedge d\psi$, *where ϕ is a p-form.*

- $d^2 = 0$.

- *If f is a function, then df is defined as in Section 6 of Chapter 5.*

We briefly mention the meaning of these properties. The fourth property says that d of a function is its *total differential*. In particular d of a constant is 0. The first property is the sum rule for derivatives; the second property generalizes the

Figure 6.4. Folium of Descartes.

product rule. The property $d^2 = 0$ is a brilliant way of organizing information. For smooth functions f we have

$$\frac{\partial^2 f}{\partial x_j \partial x_k} = \frac{\partial^2 f}{\partial x_k \partial x_j}.$$

The alternating property of the wedge product combines with this equality of mixed partials to give $d^2 = 0$. It is not hard to check that these properties uniquely determine d. More interesting is that the exterior derivative affords considerable simplification to classical vector analysis. We give two simple examples. Various laws from physics such as $\mathrm{curl}(\nabla(f)) = 0$ for a function f and $\mathrm{div}(\mathrm{curl}(F)) = 0$ for a vector field F amount to special cases of $d^2 = 0$. See [**11**] for the use of differential forms in physics.

Using d, we reformulate Green's theorem. Let us denote the interior of γ as Ω and rewrite γ as the boundary $\partial \Omega$ of Ω. Then Green's theorem states that

$$(16) \qquad \qquad \int_{\partial \Omega} \omega = \int_{\Omega} d\omega.$$

In (16), ω is a 1-form, and $\partial \Omega$ is 1-dimensional. Formula (16) holds whenever ω is a $(k-1)$-form and $\partial \Omega$ is the $(k-1)$-dimensional boundary of a k-dimensional object. The result is called the general Stokes formula. This formulation includes the fundamental theorem of calculus, Green's theorem, the divergence theorem, and the classical Stokes theorem all as special cases. This general formulation also clarifies Maxwell's equations from the theory of electricity and magnetism. See [**21**] for a complete mathematical treatment of these ideas and see [**11**] for physical interpretations.

3. Goursat's proof

This section seems at first a bit technical and can be skipped on first reading; on the other hand, the argument in Goursat's proof is beautiful and elegant, and it deserves some time and effort.

We will be integrating around rectangles. First we note a special case which we can do directly and which arises in the proof.

Example 3.1. Let ∂R be the boundary of a closed rectangle R and let α, β be complex constants. Then

$$(17) \qquad \int_{\partial R} (\alpha z + \beta) dz = 0.$$

The proof is immediate: $(\alpha z + \beta)dz = d(\alpha \frac{z^2}{2} + \beta z)$ and hence its line integral over any closed path must be 0.

In the proof we use the following fact about the topology of $\mathbf{C}$. Consider a sequence R_j of closed rectangular regions of diameter d_j in $\mathbf{C}$. Suppose for all j that $R_{j+1} \subset R_j$ and also that $\lim_{n \to \infty} d_j = 0$. Then the intersection of the regions consists of a single point. This result follows from a corresponding result on the real line. Consider a sequence I_j of closed intervals in $\mathbf{R}$ of length L_j. Suppose for all j that $I_{j+1} \subset I_j$ and also that $\lim_{n \to \infty} L_j = 0$. Then the intersection of these intervals is a single point. This statement about the real line is a consequence of the completeness axiom and in fact can be taken instead of the least upper bound property as what it means for an ordered field to be complete. See [**3**, **20**] for additional discussion.

In the proof of Goursat's theorem, we create a sequence R_n of closed rectangles which satisfy the above properties. The sequence converges to a single point.

If we take Definition 1.2 of Chapter 5 for complex analytic (in which we assume f is continuously differentiable), then we can apply Green's theorem to show that $\int_\gamma f(z)dz = 0$ whenever f is complex analytic on and inside γ. Goursat discovered that the hypothesis of continuous derivatives is redundant. In this section we will prove the following statement. If f is complex differentiable in the sense (Definition 1.3 of Chapter 5) that the limit quotient exists at every point in a domain Ω, then $\int_{\partial R} f(z)dz = 0$ for every rectangle R contained in Ω. In the next paragraph we elaborate a slightly confusing point; although we integrate only along ∂R, it is required that f be complex differentiable on the full closed rectangle.

Before we state and prove the theorem of Goursat, we wish to clarify this point. Let Ω be an open set in $\mathbf{C}$ with holes, such as an annular region. Then there exist rectangles R such that the boundary ∂R lies in Ω but R does not. In order that the next theorem apply, it is crucial that the full rectangle lie within Ω. We write ∂R to denote the boundary of R, oriented positively (counterclockwise).

Theorem 3.1 (Goursat's theorem). *Let Ω be an open subset of **C**. Assume that f is complex differentiable in Ω. Let R be a rectangle in Ω. Then $\int_{\partial R} f(z)dz = 0$.*

Proof. Let S be a rectangle in Ω and consider $I = \int_{\partial S} f(z)dz$. The first step in the proof is to observe (see Figure 6.2) that the rectangle S can be divided into four rectangles S_j by bisecting each of the sides of S. We call this procedure *quadrisection*. We then have

$$(18) \qquad |I| = \left| \int_{\partial S} f(z)dz \right| = \left| \sum_{j=1}^{4} \int_{\partial S_j} f(z)dz \right| \le 4 \max \left| \int_{\partial S_j} f(z)dz \right|.$$

We will keep dividing the original rectangle in this fashion, choosing at each stage one of the four rectangles. See Figure 6.5. Of the four choices we choose j to make $|\int_{\partial S_j} f(z)dz|$ as large as possible. Let us relabel. We call the original rectangle R; we call R_n the rectangle chosen at the n-th step of quadrisection. By (18) and induction we obtain

$$(19) \qquad |I| = \left| \int_{\partial S} f(z)dz \right| \le 4^n \left| \int_{\partial R_n} f(z)dz \right|.$$

Observe also that the diameter δ_n of R_n is 2^{-n} times the diameter δ of R and that the perimeter L_n of R_n is 2^{-n} times the perimeter L of R.

By the basic theorem in real analysis mentioned above, the limit of these rectangles will be a single point z_0. We then use the complex differentiability at z_0 to write (compare the discussion after Definition 1.3 of Chapter 5)

$$f(z) = f(z_0) + (z - z_0)f'(z_0) + E(z_0, z - z_0),$$

where the error function E is small in the sense that

$$\lim_{z \to z_0} \frac{E(z_0, z - z_0)}{(z - z_0)} = 0.$$

Given $\epsilon > 0$, we have $|E(z_0, z - z_0)| \le \epsilon |z - z_0|$ when $|z - z_0|$ is sufficiently small. We can make $|z - z_0|$ small by choosing n large in (19). We can therefore write, for z near z_0,

(20)
$$\int_{\partial R_j} f(z)dz = \int_{\partial R_j} \left(f(z_0) + (z - z_0)f'(z_0) + E(z_0, z - z_0) \right) dz$$
$$= \int_{\partial R_j} f(z_0)dz + \int_{\partial R_j} (z - z_0)f'(z_0)dz + \int_{\partial R_j} E(z_0, z - z_0)dz = \int_{\partial R_j} E(z_0, z - z_0)dz.$$

The first two integrals in the second line of (20) vanish by (17) from Example 3.1. To finally prove the theorem, we observe by (19) and (20) that

$$(21) \qquad |I| \le 4^j \left| \int_{\partial R_j} f(z)dz \right| = 4^j \left| \int_{\partial R_j} E(z_0, z - z_0)dz \right|.$$

We then use the ML-inequality to obtain

$$(22) \qquad |I| \le 4^j \left| \int_{\partial R_j} E(z_0, z - z_0)dz \right| \le 4^j \epsilon |z - z_0| L_j \le 4^j \epsilon \delta_j L_j = \epsilon \delta L.$$

Since ϵ is an arbitrary positive number and δ and L are fixed, we conclude that $|I| = 0$.   □

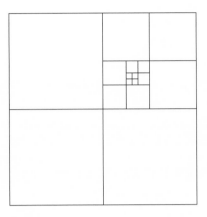

Figure 6.5. Iterated quadrisection of a rectangle.

4. The Cauchy integral formula

Let Ω be an open subset of $\mathbf{C}$. Let $f : \Omega \to \mathbf{C}$ be complex analytic on Ω. We will show, for each point $p \in \Omega$, that there is an open disk about p on which f can be developed into a convergent power series. Furthermore, the radius of the disk can be taken to be the distance from p to the boundary of Ω. We will establish this fundamental fact by using integrals. First, however, we consider the region of convergence for the function $\frac{1}{1-z}$ when we expand it around different points.

We know that $\sum_{n=0}^{\infty} z^n = \frac{1}{1-z}$ whenever $|z| < 1$; recall its generalization (5) from Chapter 4. For any $p \neq 1$ we can write, whenever $|\frac{z-p}{1-p}| < 1$,

$$(23) \qquad \frac{1}{1-z} = \frac{1}{1-p-(z-p)} = \frac{1}{1-p} \frac{1}{\frac{z-p}{1-p}} = \sum_{n=0}^{\infty} (\frac{1}{1-p})^{n+1} (z-p)^n.$$

Thus, for each p such that $\frac{1}{1-z}$ is analytic near p, there is a convergent power series representation. The radius of convergence is precisely the distance from the point p to the singular point 1. Cauchy's brilliant idea uses complex integration to reduce the general case to this particular situation. Several steps are required.

Step one is the Cauchy theorem. This result yields $\int_{\gamma} f(z)dz = 0$ for piecewise smooth simple closed curves γ when f is complex analytic on and inside the curve. We do not require the most general version.

Theorem 4.1 (Cauchy's theorem). *Let Ω be an open subset of $\mathbf{C}$ and let $f : \Omega \to \mathbf{C}$ be complex differentiable on Ω. Let γ be a piecewise smooth simple closed curve in Ω, and suppose that the interior of γ lies within Ω. Then*

$$\int_{\gamma} f(z)dz = 0.$$

Proof. We will not give a complete proof. In Theorem 3.1 we verified the conclusion whenever γ is the boundary of a rectangle. It follows that the result holds whenever γ is a simple closed curve all of whose pieces consist of line segments parallel to an axis. By taking limits, it follows that the result holds whenever γ is a limit of such curves. Such limit curves include the piecewise smooth curves. See Figure 6.3 for the geometric idea. □

We use Cauchy's theorem to derive the crucial Cauchy integral formula. This result uses integrals to show that a complex analytic function is not wildly different from a geometric series.

Theorem 4.2 (Cauchy integral formula). *Let Ω be an open set in $\mathbf{C}$, and suppose f is complex analytic on Ω. Let γ be a positively oriented piecewise smooth simple closed curve in Ω and assume that the interior of γ is contained in Ω. For each $z \in \Omega$, the following formula holds for f:*

$$(24) \qquad f(z) = \frac{1}{2\pi i} \int_\gamma \frac{f(\zeta)}{\zeta - z} d\zeta.$$

Proof. The idea of the proof is to show that the line integral on the right-hand side of (24) equals the line integral of the same integrand over each circle of sufficiently small radius ϵ about z. Then we take the limit as ϵ tends to 0, obtaining $f(z)$.

Let μ_ϵ be the circle of radius ϵ about z, traversed counterclockwise and let η_ϵ denote the curve defined by $\eta_\epsilon = \gamma - \mu_\epsilon$. The definition of line integrals over the sum of two curves yields

$$(25) \qquad \int_{\eta_\epsilon} \frac{f(\zeta)}{\zeta - z} d\zeta = \int_\gamma \frac{f(\zeta)}{\zeta - z} d\zeta - \int_{\mu_\epsilon} \frac{f(\zeta)}{\zeta - z} d\zeta.$$

We claim that the left-hand side of (25) vanishes. To see this from an intuitive fashion, connect the curves γ and μ_ϵ with a line segment ν, and then include both ν and $-\nu$ with the curve η_ϵ. Because of cancellation, including ν and $-\nu$ does not affect the integral. On the other hand (imagine an infinitesimal separation between $\pm\nu$), the new closed curve no longer winds around the singularity at z. Hence by Cauchy's theorem, applied to the complex analytic function $\zeta \to \frac{f(\zeta)}{\zeta - z}$, the line integral is 0. See Figure 6.6.

In other words, the function $\zeta \to \frac{f(\zeta)}{\zeta - z}$ is complex analytic on and inside η_ϵ (the region between the two curves). Hence the line integral is 0 by Cauchy's theorem. Therefore, for each sufficiently small ϵ we have

$$(26) \qquad \int_\gamma \frac{f(\zeta)}{\zeta - z} d\zeta = \int_{\mu_\epsilon} \frac{f(\zeta)}{\zeta - z} d\zeta.$$

Parametrizing the integral on the right-hand side of (26) using $\zeta = z + \epsilon e^{i\theta}$ gives

$$(27) \qquad \int_{\mu_\epsilon} \frac{f(\zeta)}{\zeta - z} d\zeta = \int_0^{2\pi} f(z + \epsilon e^{i\theta}) i d\theta.$$

Letting ϵ tend to 0, we obtain $2\pi i f(z)$. See Exercise 6.7. □

We could spend many pages deriving consequences of the Cauchy integral formula, but we limit ourselves to the most important basic theoretical consequences

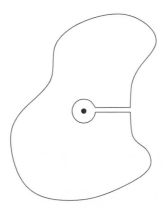

Figure 6.6. The key to the proof of the Cauchy integral formula.

and to the computation of residues. Perhaps the most important consequence is the existence of the local power series expansion. Figure 4.1 illustrates the region of convergence for the geometric series based at p. Figure 6.7 illustrates the region of the convergence of the power series for a general complex analytic function; these figures capture the essence of the proof of Theorem 4.3. The equivalence of the three possible definitions of complex analytic function also follows from Theorem 4.3. The computation of residues is appealing at this stage because of its striking application to doing calculus integrals.

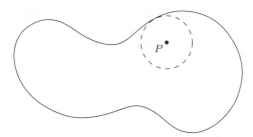

Figure 6.7. Power series and the Cauchy integral formula.

Theorem 4.3. *Suppose f satisfies (24). Then, for each $p \in \Omega$, there is a positive ϵ such that f has a convergent power series expansion on $\{z : |z - p| < \epsilon\}$. One can choose ϵ equal to any positive number less than the distance from p to the boundary of Ω.*

Proof. The student should master this proof! Choose $p \in \Omega$, and let r denote the distance from p to the boundary. This number r is positive since Ω is an open set. Choose ϵ with $0 < \epsilon < r$ and consider the circle C_ϵ of radius ϵ about p. By the Cauchy integral formula we have

$$(28) \qquad f(z) = \frac{1}{2\pi i} \int_{C_\epsilon} \frac{f(\zeta)}{\zeta - z} d\zeta.$$

Following the technique in (23), we write

$$(29) \qquad f(z) = \frac{1}{2\pi i} \int_{C_\epsilon} \frac{f(\zeta)}{\zeta - p - (z-p)} d\zeta = \frac{1}{2\pi i} \int_{C_\epsilon} \frac{f(\zeta)}{\zeta - p} \frac{1}{1 - (\frac{z-p}{\zeta-p})} d\zeta$$

$$= \frac{1}{2\pi i} \int_{C_\epsilon} \frac{f(\zeta)}{\zeta - p} \sum_{n=0}^{\infty} (\frac{z-p}{\zeta-p})^n d\zeta.$$

The geometric series converges absolutely and uniformly for $|z-p| < |\zeta-p|$. Since the circle C_ϵ and its interior lie in Ω, we have absolute and uniform convergence for $|z - p| \le \epsilon$. By Corollary 5.1 of Chapter 2 we can interchange integral and summation to obtain

$$(30) \qquad f(z) = \sum_{n=0}^{\infty} \left(\frac{1}{2\pi i} \int_{C_\epsilon} \frac{f(\zeta)}{(\zeta-p)^{n+1}} d\zeta \right) (z-p)^n = \sum_{n=0}^{\infty} c_n (z-p)^n.$$

Thus, near p, there is a convergent power series expansion for f and even an integral formula for the coefficients. $\square$

The following corollaries are immediate.

Corollary 4.1. *Under the hypotheses of Theorem 4.3, we have*

$$(31) \qquad c_n = \frac{f^{(n)}(p)}{n!} = \frac{1}{2\pi i} \int_{C_\epsilon} \frac{f(\zeta)}{(\zeta-p)^{n+1}} d\zeta.$$

Proof. The formula is immediate from (30). $\square$

Corollary 4.2 (Cauchy estimates). *Let f be complex analytic on $\{z : |z| < \rho\}$ and assume $R < \rho$. Let M_R be the maximum of $|f|$ on the circle $|z| = R$. Then*

$$|f^{(n)}(0)| \le \frac{n! M_R}{R^n}.$$

Proof. Consider the circle $|z| = R$ and apply the ML-inequality to (31). Since $L = 2\pi R$, the inequality follows. $\square$

The Cauchy estimates reveal that the successive derivatives of a complex analytic function at a point are not arbitrary complex numbers. The resulting Taylor series must be convergent, and hence some information on the size of the coefficients is forced.

Corollary 4.3. *A complex analytic function is infinitely differentiable.*

Corollary 4.4. *Suppose that all derivatives of a complex analytic function $f : \Omega \to \mathbf{C}$ vanish at p. Then $f(z) = 0$ for all z near p. If in addition Ω is connected, then $f(z) = 0$ for all $z \in \Omega$.*

In the previous corollary, we would like to say that $f(z) = 0$ for all z, but we cannot do so without assuming that f is defined on a *connected* set. For example, if Ω were the union of two disjoint balls, f could be 0 on the first ball and something else on the second ball. Then all derivatives of f would vanish at points on the first ball, but f would not be identically 0. In the next two results we assume connectedness as a hypothesis. Although our discussion from Section 6 of Chapter

1 on connectedness was brief, the concept is intuitive and no difficulties arise. We mention that a ball is connected; in most applications one may work in a ball. We sketch the proof of the second statement from Corollary 4.4. Consider the subset A of Ω consisting of points near which f vanishes identically. By the first part of the corollary, this set is open. On the other hand, the set B of points in Ω near which f does not vanish identically is also open. Then we have $A \cup B = \Omega$. If Ω is connected, then one of these sets must be empty.

The following theorem also follows from the Cauchy integral formula. It reveals one sense in which complex analytic functions behave like polynomials. The integer m in this result is called the *order* of the zero of f at L.

Theorem 4.4. *Suppose f is complex analytic in a connected open set Ω containing p and $f(p) = 0$. Then either $f(z) = 0$ for all z in Ω or there is a positive integer m and a complex analytic function g on Ω such that*

- $g(p) \neq 0$,
- $f(z) = (z - p)^m g(z)$ *for all z in Ω.*

Proof. Expand f in a power series about p. Either all the Taylor coefficients at p are 0, in which case Corollary 4.4 yields $f(z) = 0$ for all z in Ω, or there is some smallest m for which $c_m \neq 0$. In the second case we define g by the rule $g(p) = c_m$ and $g(z) = \frac{f(z)}{(z-p)^m}$ for $z \neq p$. Both conclusions hold. □

Corollary 4.5. *Suppose f is complex analytic on an open connected set Ω. Let $\{a_n\}$ be a sequence of distinct points in Ω which converges to a limit p in Ω, and suppose $f(a_n) = 0$ for all n. Then $f(z) = 0$ for all z in Ω.*

Proof. If f is not identically zero, then we can write $f(z) = (z - p)^m g(z)$ as in the previous theorem. In particular $g(p) \neq 0$. But $f(a_n) = 0$ for all n implies $g(a_n) = 0$ for all n, and hence by continuity $g(p) = 0$. The only possibility is therefore that f is identically zero. □

In Theorem 4.4 we assumed that the a_n are distinct. We could have made a weaker assumption, as long as we avoid the following problem. The conclusion fails for example if $a_n = p$ for all large n; the equation $f(a_n) = (a_n - p)^m g(a_n)$ does not then allow us to conclude that $g(a_n) = 0$. The next remark is a bit more subtle.

Remark 4.1. The previous corollary requires that the limit point p lie in Ω. Here is a counterexample without that assumption. Let Ω be the complement of the origin, and let $f(z) = \sin(\frac{\pi}{z})$. Then $f(\frac{1}{n}) = 0$ for all n but f is not identically zero. Note that f is not complex analytic at 0.

▶ **Exercise 6.10.** Assume f is continuous at z. Let μ_ϵ be the positively oriented circle of radius ϵ about z. Prove that

$$\lim_{\epsilon \to 0} \int_{\mu_\epsilon} \frac{f(z)}{\zeta - z} d\zeta = 2\pi i f(z).$$

The next result illustrates the Cauchy integral formula, but it holds in a more general situation. We employ it in our work on doing real integrals using complex variables. We assume that the circle is positively oriented. The reader should pause and understand what it means for the series in Proposition 4.1 to converge, as both limits of summation are infinite.

Proposition 4.1. *Suppose that $\sum_{n=-\infty}^{\infty} a_n(z-p)^n = f(z)$ for $0 < |z-p| < R$, where the series converges absolutely and uniformly on each closed subannulus $\delta \leq |z| \leq r$. Then for $\epsilon < R$ we have*

$$(32) \qquad \int_{|z-p|=\epsilon} f(z)dz = 2\pi i a_{-1}.$$

Proof. We can interchange the order of integral and summation. The result then follows from the specific evaluations; for $n \neq -1$, we have

$$\int_{|z-p|=\epsilon} (z-p)^n dz = 0$$

and for $n = -1$ the integral is $2\pi i$. $\qquad\qquad\qquad\qquad\qquad\qquad\square$

Definition 4.1. Assume $\sum_{n=-\infty}^{\infty} a_n(z-p)^n = f(z)$ converges for $0 < |z-p| < R$. The number a_{-1}, namely the coefficient of $\frac{1}{z-p}$ in the expansion, is called the *residue* of f at p.

This proposition allows us to evaluate certain integrals in our heads. Given f, we need only find residues. In Chapter 7 we apply this idea to evaluate many nontrivial definite integrals from calculus.

5. A return to the definition of complex analytic function

This section unifies much of what we have done by proving that three possible definitions of *complex analytic function* are equivalent.

Theorem 5.1. *Let Ω be an open subset of $\mathbf{C}$ and suppose $f : \Omega \to \mathbf{C}$. The following three statements are equivalent:*

1) *For all $p \in \Omega$, there is a disk about p, lying in Ω, on which f can be developed in a convergent power series:*

$$(33) \qquad f(z) = \sum_{n=0}^{\infty} a_n(z-p)^n.$$

2) *The function f is continuously differentiable, and for all $p \in \Omega$,*

$$(34) \qquad \frac{\partial f}{\partial \bar{z}}(p) = 0.$$

3) *For all $p \in \Omega$, f has a complex derivative $f'(p)$, defined by the existence of the limit in (35):*

$$(35) \qquad f'(p) = \lim_{\zeta \to 0} \frac{f(p+\zeta) - f(p)}{\zeta}.$$

Proof. Assume 1) holds. Fix $p \in \Omega$ and consider a disk about p on which f has a convergent power series expansion. By Theorem 2.1 of Chapter 4, the series converges absolutely and uniformly on any closed subdisk and hence defines a continuously differentiable function. Further we may differentiate the series term by term. But $\frac{\partial}{\partial \bar{z}}((z-p)^n) = 0$ for all n. Hence $\frac{\partial f}{\partial \bar{z}} = 0$ on Ω and therefore 2) holds.

Furthermore, if 1) holds, then we may also differentiate the series term by term with respect to z. Since the radius of convergence of the derived series is the same as that of the original series, $f'(p)$ exists at each point of Ω. Thus 3) holds.

Assume 2) holds. Then we may apply Green's theorem as in Section 2 to conclude that Cauchy's theorem holds. Hence the Cauchy integral formula holds. We used the Cauchy integral formula to derive the power series expansion. Hence 2) implies 1).

It remains to prove that 3) implies either 1) or 2). Assume that 3) holds. By Goursat's proof, the line integral of $f(z)dz$ over the boundary of a rectangle equals 0 for any rectangle in Ω. Therefore the line integral vanishes over more general closed curves, Cauchy's theorem holds, and hence the Cauchy integral formula holds. From it we obtain 1). $\qquad\square$

▶ **Exercise 6.11.** Show directly that the Cauchy integral formula holds if and only if

$$\int_\gamma \frac{f(\zeta) - f(z)}{\zeta - z} d\zeta = 0.$$

▶ **Exercise 6.12.** Assume f is complex analytic for z near ζ. For $z \neq \zeta$, define $g(z)$ by

$$g(z) = \frac{f(z) - f(\zeta)}{z - \zeta}.$$

What value must $g(\zeta)$ be in order that g be complex analytic at ζ?

The next three exercises illustrate the usefulness of Theorem 5.1. By choosing the right characterization, one can find the most natural and direct proof.

▶ **Exercise 6.13.** Assume f, g are complex analytic on Ω. Then $f + g$ and fg are also complex analytic on Ω. Prove both statements using each of the three possible definitions of complex analytic function.

▶ **Exercise 6.14.** Assume f is complex analytic on Ω. Then f' also is complex analytic there. Prove this statement using each of the three possible definitions of complex analytic function. Comment: The proof using convergent power series is the only one with any subtlety.

▶ **Exercise 6.15.** Suppose $f(z) = \sum_{n=0}^{\infty} a_n(z-p)^n$, where the series converges for z near p. Assume $a_0 \neq 0$. Then $\frac{1}{f}$ is complex analytic near p. Prove this statement using each of the three definitions of complex analytic.

▶ **Exercise 6.16.** Suppose $\sum_{n=0}^{\infty} a_n z^n$ is a formal power series and $a_0 \neq 0$. Show that there is a formal power series equal to its reciprocal.

▶ **Exercise 6.17.** Let γ be a simple closed curve in the region from Figure 1.8. Determine all possible values for $\int_\gamma \frac{dz}{z-p}$.

Chapter 7

Applications of Complex Integration

In this chapter we provide applications of the Cauchy integral formula to calculus problems. Given a definite integral over an interval as in elementary calculus, we create a simple closed curve γ in $\mathbf{C}$ including this interval as one of its pieces. The techniques of Chapter 6 enable us to evaluate the line integral over γ, often by inspection or easy computation. If we can understand the behavior on the extra pieces of the contour, often in a limiting case, then we can solve the original problem. We provide many examples and several general results concerning this technique. In this chapter we also introduce the Fourier transform, the Gaussian probability distribution, and the Gamma function.

1. Singularities and residues

Suppose f is complex analytic on the set $0 < |z - p| < \epsilon$. What can happen at p itself? There are three possibilities. The first, called a *removable singularity*, arises when f is in fact also analytic at p, or when there is a value for $f(p)$ making f analytic on the set $|z - p| < \epsilon$. The following two examples illustrate this situation:

$$(1) \qquad\qquad f(z) = \frac{z^2 - p^2}{z - p},$$

$$(2) \qquad\qquad f(z) = \frac{\sin(z)}{z}.$$

In (1) the complex analytic function F defined by $F(z) = z + p$ agrees with f when $0 < |z - p|$. In (2), where p is the origin, we can extend f to a complex analytic function F, defined on all of $\mathbf{C}$. Note that $F(0) = 1$. The function F has an explicit power series expansion valid in all of $\mathbf{C}$. Riemann's removable singularities theorem, which we do not require, gives a simple test for a singularity being removable: if f

is complex analytic near a singularity at p and f is bounded, then the singularity is removable.

The second possible kind of singularity is called a *pole*. In this case, there is a (smallest) positive integer k such that $(z-p)^k f(z)$ has a removable singularity at p. We say that f has a pole of order k at p. When $k = 1$, we sometimes say that f has a *simple pole* at p. A removable singularity can be regarded as a pole of order zero. Here are two examples of poles:

$$(3) \qquad\qquad f(z) = \frac{e^z}{z^3},$$

$$(4) \qquad\qquad f(z) = \frac{1}{z^2 + 1}.$$

In (3), f has a pole of order 3 at $p = 0$. In (4), f has simple poles at $\pm i$.

The third kind of singularity includes all the remaining possibilities: f has a singularity at p, the singularity is not removable, and there is no k for which f has a pole of order k. The resulting situation is called an *essential singularity*. Here are two examples of essential singularities at 0 (see also Remark 4.1 of Chapter 6):

$$(5) \qquad\qquad f(z) = e^{\frac{1}{z^2}},$$

$$(6) \qquad\qquad f(z) = \sin(\frac{1}{z}).$$

We next discuss techniques for finding residues. Let us first recall why we care. In the previous chapter we briefly considered functions f for which $f(z) = \sum_{n=-\infty}^{\infty} a_n(z-p)^n$ on the annulus $0 < |z-p| < R$. The series here is called a *Laurent series*; it converges absolutely and uniformly on each closed subannulus $\delta \leq |z| \leq r$. In our applications the existence of the Laurent series is evident; hence we do not prove a general statement. Our main concern is the following result from Chapter 6. For $\epsilon < R$ we have

$$\int_{|z-p|=\epsilon} f(z)dz = 2\pi i a_{-1}.$$

We need only find a_{-1} in order to evaluate the integral.

Suppose f is complex analytic on the set $0 < |z-p| < \epsilon$. In case f has a removable singularity at p, the residue of f at p is zero. The converse assertion fails; for example, if $f(z) = \frac{1}{z^2}$, then f has a pole of order 2 at 0, but the residue equals 0, because there is no $\frac{1}{z}$ term in the expansion. Suppose that f is known to have a pole at p. We can find the residue there by the following result.

Lemma 1.1. *Suppose that f has a pole of order k at p. Then the residue of f at p has the following value:*

$$(7) \qquad\qquad a_{-1} = \frac{1}{(k-1)!}(\frac{\partial}{\partial z})^{k-1}\left((z-p)^k f(z)\right)(p).$$

Proof. By definition of a pole, the function $g(z) = f(z)(z-p)^k$ is complex analytic near p and hence can be expanded in a Taylor series in powers of $(z-p)$. Near p

we can write

$$g(z) = \sum_{n=0}^{\infty} b_n(z-p)^n = \sum_{n=0}^{k-1} b_n(z-p)^n + h(z)(z-p)^k$$

where h is complex analytic. Away from p we then have

$$(8) \qquad f(z) = \frac{b_0}{(z-p)^k} + ... + \frac{b_{k-1}}{(z-p)} + h(z).$$

Thus the residue is b_{k-1}, namely the $(k-1)$-st derivative of g at p divided by $(k-1)!$, and (7) follows. $\qquad\square$

In practice we are often given an expression where f appears in the denominator and f has a simple zero at p. We can find the residue, with little computation, as follows.

Lemma 1.2. *Suppose f and g are complex analytic near p and f has a zero of first order there. Then the residue of $\frac{g}{f}$ at p equals*

$$\lim_{z \to p} \left(\frac{(z-p)g(z)}{f(z)} \right) = \frac{g(p)}{f'(p)}.$$

Proof. The two expressions are equal by l'Hospital's rule. We verify that the first equals the residue. If $f(z) = c_1(z-p) + c_2(z-p)^2 + ...$, then $f'(p) = c_1$. Using Theorem 4.4 from Chapter 6 and Exercise 6.15, we get

$$\frac{g(z)}{f(z)} = \frac{g(z)}{(z-p)(c_1 + ...)} = \frac{g(z)}{c_1(z-p)} (1 + ...),$$

where the ... terms are divisible by $z-p$. The result now follows from the definition of the residue at p. $\qquad\square$

Example 1.1. Put $f(z) = \frac{e^{\pi z}}{z^2+1}$. The residue of f at i is $\frac{e^{i\pi}}{2i} = \frac{i}{2}$.

2. Evaluating real integrals using complex variables methods

The simplest sort of example involves integrals over the interval $[0, 2\pi]$ involving cosine and sine. We reduce these integrals to line integrals over the unit circle and we evaluate the resulting line integral using residues.

Example 2.1. Consider the integral

$$J = \int_0^{2\pi} \frac{d\theta}{\frac{17}{4} + 2\cos(\theta)}.$$

To evaluate J, we set $z = e^{i\theta}$ and convert J into a line integral over $|z| = 1$. We have $dz = ie^{i\theta}d\theta = izd\theta$ and hence we get

$$J = \int_{|z|=1} \frac{dz}{iz(\frac{17}{4} + z + \frac{1}{z})} = \frac{1}{i} \int_{|z|=1} \frac{dz}{(z + \frac{1}{4})(z + 4)}.$$

Now the evaluation is simple. The integrand is complex analytic except at $\frac{-1}{4}$ and -4. Of these points, only $\frac{-1}{4}$ is inside the unit circle. The residue there is $\frac{1}{\frac{-1}{4}+4}$. Hence we obtain

$$J = 2\pi i \text{ Residue} = 2\pi i \frac{1}{i} \frac{4}{15} = \frac{8\pi}{15}.$$

We next consider a standard sort of example where we need to introduce extra contours to create a simple closed curve. From calculus one can evaluate

$$I = \int_{-\infty}^{\infty} \frac{dx}{x^2 + 1}$$

as follows. By definition of improper integral (here the limits are infinite) and by symmetry, we have

$$I = \lim_{R \to \infty} \int_{-R}^{R} \frac{dx}{x^2 + 1} = \lim_{R \to \infty} \left(\tan^{-1}(R) - \tan^{-1}(-R)\right) = \frac{\pi}{2} - \frac{-\pi}{2} = \pi.$$

We next use complex variables to provide an alternative evaluation of I.

Example 2.2. We evaluate I. Consider the curve γ_R defined to be the interval along the real axis from $-R$ to R together with the semi-circle η_R of radius R in the upper half-plane, traversed counterclockwise, making γ_R into a simple closed positively oriented curve. See Figure 7.1.

Assume that $R > 1$. To compute $\int_{\gamma_R} \frac{dz}{z^2+1}$, we first observe that the function $z \to \frac{1}{z^2+1}$ is complex analytic in all of $\mathbf{C}$ except for the singularities at $\pm i$. Hence it is analytic on and inside γ_R, except for the single point i. The residue at i equals $\frac{1}{2i}$. Either by applying the Cauchy integral formula to $f(z) = \frac{1}{z+i}$ or by applying Proposition 4.1 of Chapter 6, we obtain

$$\int_{\gamma_R} \frac{dz}{z^2+1} = \int_{\gamma_R} \frac{dz}{(z-i)(z+i)} = 2\pi i \frac{1}{i+i} = \pi.$$

This result holds for all R greater than 1. We let R tend to infinity. If we can show that the integral along the circular arc η_R tends to 0, then we obtain the result $I = \pi$. To show that the integral J_R along the circular arc η_R tends to 0 is not difficult in this case. Put $z = Re^{i\theta}$ and use the ML-inequality. For some constant C we get

$$(9) \qquad |J_R| = \left| \int_{\eta_R} \frac{dz}{z^2+1} \right| = \left| \int_0^{2\pi} \frac{Rie^{i\theta}d\theta}{R^2 e^{2i\theta} + 1} \right| \le \frac{C}{R}.$$

It follows from (9) that J_R tends to 0 as required.

▶ **Exercise 7.1.** Verify the inequality in (9).

The method from this example leads to the following general result, whose proof so closely follows the example that it can be safely left to the reader.

Theorem 2.1. *Let* $f(z) = \frac{p(z)}{q(z)}$ *be a rational function where the following two additional statements hold. First,* $q(z) \ne 0$ *on the real axis. Second, the degree of*

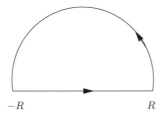

$-R$ R

Figure 7.1. Contour for Theorem 2.1.

q *is at least two more then the degree of p. Let Σ denote the sum of the residues in the upper half-plane. Then*

$$\int_{-\infty}^{\infty} f(x)dx = 2\pi i \Sigma.$$

▶ **Exercise 7.2.** Prove Theorem 2.1.

Next we give a rather amazing method for evaluating certain integrals over the positive real axis. Consider

$$I = \int_{0}^{\infty} \frac{dx}{q(x)}$$

where q is a polynomial of degree at least two, and $q(x) > 0$ for $x \geq 0$. If q were *even*, then the value of I would be half the integral over the whole real line and hence could be evaluated by the method of the previous theorem. In general we can find I by introducing logarithms.

Consider the branch of the logarithm defined by $\log(z) = \log(|z|) + i\theta$, where $0 < \theta < 2\pi$. We will integrate on the positive real axis twice, using $\theta = 0$ once and then using $\theta = 2\pi$ the second time, as suggested by the contour $\gamma_{\epsilon,R}$ from Figure 7.2. We consider the line integral

$$J_{\epsilon,R} = \int_{\gamma_{\epsilon,R}} \frac{\log(z)dz}{q(z)}.$$

For R sufficiently large and ϵ sufficiently small, the value of $J_{\epsilon,R}$ is $2\pi i$ times the sum of the residues of $\left(\frac{\log(z)}{q(z)}\right)$ in the entire complex plane. In the proof of Theorem 2.2 we show that the integrals on the circular arcs tend to 0 as ϵ tends to 0 and R tends to ∞. What happens on the positive real axis? Note that $\log(z) = \log(|z|)$ on the top part, but $\log(z) = \log(|z|) + 2\pi i$ on the bottom part. The bottom part is traversed in the opposite direction from the top part; hence everything cancels except for the term

$$\int_{R}^{\epsilon} \frac{2\pi i \, dz}{q(z)}.$$

After canceling out the common factor of $2\pi i$ and including the residues in all of **C** in the sum, we find

(10) $$I = -\sum \text{residues} \left(\frac{\log(z)}{q(z)}\right).$$

We state a simple generalization of this calculation as the next theorem.

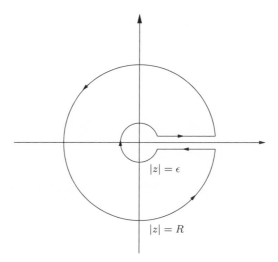

Figure 7.2. Contour for Theorem 2.2.

Theorem 2.2. *Let $\frac{f}{q}$ be a rational function, where the degree of q is at least two more than the degree of f, and suppose that $q(x) \neq 0$ for $x \geq 0$. Then the integral in (11) converges and*

(11)
$$\int_0^\infty \frac{f(x)}{q(x)} dx = -\sum \text{residues} \left(\frac{\log(z)f(z)}{q(z)} \right).$$

The sum in (11) is taken over all roots of q in $\mathbf{C}$.

Proof. (Sketch) We proceed as in the discussion above, using the contour given in Figure 7.2. Let $J = J_{\epsilon,R}$ denote the integral

$$\int_{\gamma_{\epsilon,R}} \frac{f(z)\log(z)}{q(z)} dz.$$

For sufficiently small ϵ and sufficiently large R we have $J = 2\pi i \Sigma$, where Σ is the sum of the residues. As above, the integrals along the top and bottom parts of the real axis cancel except for the factor $2\pi i \int_R^0 \frac{f(x)}{q(x)} dx$. If we can show that the integrals along the circular arcs tend to 0 as ϵ tends to 0 and R tends to infinity, then (11) follows. Thus the main point is to establish these limits.

For $|z| = \epsilon$, the integral becomes

(12)
$$\int_0^{2\pi} \frac{q(\epsilon e^{i\theta})}{f(\epsilon e^{i\theta})} \epsilon i e^{i\theta} \left(\log(\epsilon) + i\theta \right) d\theta.$$

Note that f and q are continuous at 0 and $q(0) \neq 0$. Since $\lim_{\epsilon \to 0} \epsilon \log(\epsilon) = 0$, it follows that (12) tends to 0 as ϵ tends to 0. Next replace ϵ by R in (12). By assumption, the degree of q is at least two larger than the degree of f. Hence, for R large enough, there is a constant C such that we can estimate $|\frac{f(Re^{i\theta})}{q(Re^{i\theta})}|$ by $\frac{C}{R^2}$. It follows that the entire integral can be estimated by $\frac{C_1 + C_2 \log(R)}{R}$. The limit of

$\frac{\log(R)}{R}$ as R tends to infinity is zero. It follows that the entire integral tends to 0 as well. Formula (11) follows. □

Example 2.3. First we compute

$$I = \int_0^\infty \frac{dx}{x^2 + 1}$$

by this method. There are two singularities. Using (11), we obtain

(13) $$I = -\left(\frac{\log(i)}{2i} + \frac{\log(-i)}{-2i}\right) = -\left(\frac{\pi}{4} - \frac{3\pi}{4}\right) = \frac{\pi}{2}.$$

Example 2.3 can be done by Theorem 2.1, but the next example cannot.

Example 2.4. Next we consider $p(x) = x^2 + x + 1$. We wish to find

$$I = \int_0^\infty \frac{dx}{x^2 + x + 1}.$$

Notice that $z^3 - 1 = (z - 1)(z^2 + z + 1)$, and hence the poles are at the complex cube roots of unity ω and $\overline{\omega}$. Note that $2\omega + 1 = i\sqrt{3}$ and $2\overline{\omega} + 1 = -i\sqrt{3}$. By (11) we obtain

(14) $$I = -\left(\frac{\log(\omega)}{2\omega + 1} + \frac{\log(\overline{\omega})}{2\overline{\omega} + 1}\right) = -\left(\frac{2\pi}{3\sqrt{3}} - \frac{4\pi}{3\sqrt{3}}\right) = \frac{2\pi}{3\sqrt{3}}.$$

Remark 2.1. The same technique can be used to evaluate integrals involving powers of the logarithm. One includes one extra logarithm and proceeds as in the next several exercises.

▶ **Exercise 7.3.** Assume $a > 0$ and n is a positive integer. Choose the branch of logarithm from Theorem 2.2. Find the residue of $\frac{(\log(z))^n}{(z+a)^2}$ at $z = -a$.

▶ **Exercise 7.4.** Assume $a > 0$. The purpose of this exercise is to evaluate

$$I = \int_0^\infty \frac{\log(x)}{(x + a)^2} dx.$$

The evaluation generalizes the method of proof from Theorem 2.2. Use the same contour, but this time consider the integral

$$\int_{\gamma_{\epsilon, R}} \frac{(\log(z))^2}{(z + a)^2} dz.$$

On the bottom part of the real axis, we now get the term $(\log(z) + 2\pi i)^2$. Expanding the square shows that we now get three terms, and only the first term cancels the term on the top part. Fill in the details to verify that $I = \frac{\log(a)}{a}$. Note: The pole at $-a$ is of order two. Use the previous exercise to find the residue there.

▶ **Exercise 7.5.** (Difficult) Find

$$I = \int_0^\infty \frac{(\log(x))^n}{(x + 1)^2} dx.$$

We next compute an important integral where estimation on a circular arc is rather subtle. The technique used in this example often arises in harmonic analysis. We start with three simple facts about the sine function.

Lemma 2.1. *There is a positive constant C such that $\frac{\sin(\theta)}{\theta} \geq C$ for $0 \leq \theta \leq \frac{\pi}{2}$.*

Proof. Define g by $g(\theta) = \frac{\sin(\theta)}{\theta}$ for $\theta \neq 0$ and $g(0) = 1$. Then g is continuous on $[0, \frac{\pi}{2}]$ and $g(\theta) \neq 0$. By basic analysis (Theorem 1.1 of Chapter 8), g achieves a positive minimum value C. $\qquad\square$

Lemma 2.2. *For $0 \leq \theta \leq \frac{\pi}{2}$ and $R > 0$ there is a positive number K such that*

$$(15) \qquad \int_0^{\frac{\pi}{2}} e^{-R\sin(\theta)}\,d\theta \leq \int_0^{\frac{\pi}{2}} e^{-RC\theta}\,d\theta \leq \frac{K}{R}.$$

Proof. By Lemma 2.1 we have $\sin(\theta) \geq C\theta$ and hence $R\sin(\theta) \geq RC\theta$. Since multiplying by -1 switches an inequality and exponentiating preserves one, we get

$$(16) \qquad\qquad e^{-R\sin(\theta)} \leq e^{-RC\theta}.$$

Estimating the integral from the middle term in (15) by using (16), we obtain a constant times the factor of $\frac{1}{R}$. $\qquad\square$

Lemma 2.3.

$$\int_0^{\pi} e^{-R\sin(\theta)}\,d\theta = 2\int_0^{\frac{\pi}{2}} e^{-R\sin(\theta)}\,d\theta.$$

Proof. Obvious! $\qquad\square$

Example 2.5. We evaluate the integral

$$I = \int_{-\infty}^{\infty} \frac{x\sin(x)}{x^2 + 1}\,dx = \frac{\pi}{e}.$$

Note that

$$I = \text{Im}\left(\int_{-\infty}^{\infty} \frac{xe^{ix}}{x^2 + 1}\,dx\right).$$

We use the usual circular contour from Theorem 2.1. The hard part is to show that the integral over the semi-circle tends to 0 as R tends to infinity. We postpone checking that fact. Assuming it is true, we obtain

$$I = \text{Im}\left((2\pi i)\text{Residue}_i\left(\frac{ze^{iz}}{z^2 + 1}\right)\right) = \text{Im}\left((2\pi i)\left(\frac{ie^{-1}}{2i}\right)\right) = \frac{\pi}{e}.$$

Next we investigate the integral I_R along the semi-circle $|z| = R$, where we assume $R > 1$. We have

$$I_R = \int_0^{\pi} \frac{Re^{Re^{i\theta}}}{R^2 e^{2i\theta} + 1} Rie^{i\theta}\,d\theta = \int_0^{\pi} \frac{Re^{Ri\cos(\theta) - R\sin(\theta)}}{R^2 e^{2i\theta} + 1} Rie^{i\theta}\,d\theta.$$

Note that $|e^{Ri\cos(\theta)}| = 1$. Also note that $|\frac{1}{R^2 e^{2i\theta}+1}| \leq \frac{1}{R^2 - 1}$. We use the three previous lemmas and Lemma 1.1 of Chapter 6 to estimate the integral as follows:

$$I_R \leq \frac{R^2}{R^2 - 1}\int_0^{\pi} e^{-R\sin(\theta)}\,d\theta \leq \frac{2R^2}{R^2 - 1}\int_0^{\frac{\pi}{2}} e^{-R\sin(\theta)}\,d\theta$$

$$\leq c'\int_0^{\frac{\pi}{2}} e^{-RC\theta}\,d\theta \leq \frac{c}{R}.$$

Here c' and c are unimportant positive constants. Hence the integral along the semi-circle tends to 0 as R tends to infinity.

Example 2.6. We evaluate the integrals $I_c = \int_0^\infty \cos(x^2)dx$ and $I_s = \int_0^\infty \sin(x^2)dx$. Each turns out to equal $\sqrt{\frac{\pi}{8}}$.

We consider the integral of e^{-z^2} around the simple closed curve γ_R consisting of three pieces: the interval $[0, R]$ on **R**, the arc of the circle $|z| = R$ for $0 \leq \theta \leq \frac{\pi}{4}$, and the forty-five degree line segment from $Re^{\frac{i\pi}{4}}$ to 0. By Cauchy's theorem $\int_{\gamma_R} e^{-z^2}dz = 0$. Exercise 7.6 asks you to check that the integral along the semi-circular arc tends to 0 as R tends to infinity. We parametrize the forty-five degree line segment by $z = e^{\frac{i\pi}{4}}t$ for $R \geq t \geq 0$. Hence we obtain

$$(17) \qquad \int_0^\infty e^{-x^2}dx = \int_0^\infty e^{-it^2}e^{\frac{i\pi}{4}}dt = (\frac{1+i}{\sqrt{2}})\int_0^\infty e^{-it^2}dt.$$

Next we substitute $e^{-it^2} = \cos(t^2) - i\sin(t^2)$ into (17). By Proposition 3.1 of the next section, the left-hand side of (17) equals $\frac{\sqrt{\pi}}{2}$. Equating imaginary parts of both sides of (17) yields $I_c = I_s$. Equating real parts of both sides of (17) now yields

$$\frac{\sqrt{\pi}}{2} = \frac{1}{\sqrt{2}}(I_c + I_s) = \frac{2}{\sqrt{2}}I_c,$$

from which the conclusion follows.

▶ **Exercise 7.6.** Fill in the details of the evaluations and estimations in Examples 2.5 and 2.6.

▶ **Exercise 7.7.** Evaluate the following integrals using the techniques in this chapter. (Some answers are given.)

1) For $a > 0$, show that

$$\int_{-\infty}^\infty \frac{dx}{x^2 + a^2} = \frac{\pi}{a}.$$

2) For $a, b > 0$, show that

$$\int_{-\infty}^\infty \frac{dx}{(x^2 + a^2)(x^2 + b^2)} = \frac{\pi}{ab(a + b)}.$$

3) For $a > 0$ and $n \in \mathbf{N}$, find

$$\int_{-\infty}^\infty \frac{dx}{(x^2 + a^2)^n}.$$

4) For $k \in \mathbf{N}$, show that

$$\int_0^{2\pi} \sin^{2k}(\theta)d\theta = \frac{\pi}{2^{2k-1}}\binom{2k}{k}.$$

5) For $j, k \in \mathbf{N}$, find

$$\int_0^{2\pi} \cos^{2j}(\theta)\sin^{2k}(\theta)d\theta.$$

6) For $a, b > 0$, show that

$$\int_0^{2\pi} \frac{d\theta}{a^2\cos^2(\theta) + b^2\sin^2(\theta)} = \frac{2\pi}{ab}.$$

7) For $\alpha > 1$, show that

$$\int_0^\infty \frac{dx}{1+x^\alpha} = \frac{\frac{\pi}{\alpha}}{\sin(\frac{\pi}{\alpha})}.$$

8) Use Theorem 2.2 to show that

$$\int_0^\infty \frac{dx}{(x^2+1)(x+1)} = \frac{\pi}{4}.$$

9) For p an integer with $p > 2$, let I_p be defined as follows:

$$I_p = \int_0^\infty \frac{dx}{1+x+x^2+...+x^{p-1}}.$$

   a) Find the limit of I_p as p tends to infinity. (The limit is not hard to find.)

   b) Prove that

$$I_p = \frac{2\pi}{p^2} \sum_{j=1}^{p-1} j\left(\sin(\frac{4\pi j}{p}) - \sin(\frac{2\pi j}{p})\right).$$

10) Show that

$$\int_{-\infty}^\infty \frac{\sin^2(x)}{x^2} dx = \pi.$$

3. Fourier transforms

The Fourier transform is an amazing tool in both pure and applied mathematics. One reason behind its success is that it converts differentiation (a difficult idea) into multiplication (an easier idea). We ask the reader to ponder the following question before reading the subsequent development: what would the half-derivative of a function be? We will give an elegant answer.

The first item of business is the following result about integrals.

Proposition 3.1. *For all real y, the value of the integral I_y in (18) is $\sqrt{\pi}$.*

(18) $$I_y = \int_{-\infty}^\infty e^{-(x+iy)^2} dx = \sqrt{\pi}.$$

Proof. First we establish the independence of the integral on y. Let γ_R denote the rectangle with vertices at $-R, R, R+iy, -R+iy$. We traverse γ_R counterclockwise as usual. Since $z \to e^{-z^2}$ is complex analytic in all of $\mathbf{C}$, Cauchy's theorem yields

$$0 = \int_{\gamma_R} e^{-z^2} dz.$$

Breaking up this line integral into four pieces yields

(19) $0 = \int_{-R}^R e^{-x^2} dx + \int_0^y e^{-(R+it)^2} i\, dt + \int_R^{-R} e^{-(x+iy)^2} dx + \int_y^0 e^{-(-R+it)^2} i\, dt.$

We let R tend to infinity in (19). If we can show that the limits of the second and fourth terms are 0, we obtain

$$0 = \int_{-\infty}^{\infty} e^{-x^2}\,dx - \int_{-\infty}^{\infty} e^{-(x+iy)^2}\,dx,$$

and the independence of I_y on y then follows.

It remains to estimate the integrals on the vertical segments. They are almost identical; we estimate one of them:

$$(20) \qquad \left| \int_0^y e^{-(R+it)^2}\,dt \right| \le \int_0^y |e^{-(R+it)^2}|\,dt = \int_0^y e^{-R^2} e^{t^2}\,dt \le e^{y^2} e^{-R^2}.$$

Since y is fixed and $\lim_{R\to\infty} e^{-R^2} = 0$, we obtain the needed limit.

Therefore it suffices to evaluate (18) when $y = 0$. The integral, whose value is $\sqrt{\pi}$, cannot be done by the methods of one-variable calculus. One standard method of evaluating it, which we generalize considerably in the next section, is to consider its square and to use polar coordinates in the plane. Let I denote the integral, consider I^2, and note that now y is a dummy variable. We obtain

$$I^2 = \int_{-\infty}^{\infty}\int_{-\infty}^{\infty} e^{-(x^2+y^2)}\,dx\,dy = \int_0^{\infty}\int_0^{2\pi} e^{-r^2} r\,d\theta\,dr = \frac{1}{2} 2\pi = \pi.$$

$\square$

Definition 3.1. Let $f : \mathbf{R} \to \mathbf{C}$ be an integrable function. Its Fourier transform, denoted by $\hat{f}$, is defined by

$$(21) \qquad \hat{f}(\xi) = \frac{1}{\sqrt{2\pi}} \int_{-\infty}^{\infty} f(x) e^{-ix\xi}\,dx.$$

We will consider some technical points about this definition, but our investigations will not get heavily involved in these subtleties. First of all, when we say that f is integrable, we mean that the integral of $|f|$ exists. Second, the integral in (21) is *improper*, because both limits of integration are infinite. Therefore the integral notation in (21) is an abbreviation for a double limit:

$$\int_{-\infty}^{\infty} f(x) e^{-ix\xi}\,dx = \lim_{a\to-\infty}\lim_{b\to\infty} \int_a^b f(x) e^{-ix\xi}\,dx.$$

In some cases this double limit fails to exist but the limit

$$\int_{-\infty}^{\infty} f(x) e^{-ix\xi}\,dx = \lim_{R\to\infty} \int_{-R}^{R} f(x) e^{-ix\xi}\,dx$$

does exist. Another interesting point concerns extensions of the definition to more general functions or even to *generalized functions*, or *distributions*. Once this sophisticated theory has been developed, it is possible to take Fourier transforms of more general functions and distributions. Often one writes the same integral, but a more subtle procedure has in fact been used.

Theorem 3.1. *Put* $f(x) = e^{\frac{-x^2}{2a^2}}$. *Then the Fourier transform* $\hat{f}$ *is given by*

$$(22) \qquad \hat{f}(\xi) = a e^{\frac{-\xi^2 a^2}{2}}.$$

In particular, if $g(x) = e^{\frac{-x^2}{2}}$, *then* $\hat{g} = g$.

Proof. We must evaluate the integral

$$I = \frac{1}{\sqrt{2\pi}} \int_{-\infty}^{\infty} e^{\frac{-x^2}{2a^2}} e^{-ix\xi} dx.$$

First we put $x = ay$ to get

$$I = \frac{a}{\sqrt{2\pi}} \int_{-\infty}^{\infty} e^{\frac{-y^2}{2}} e^{-iya\xi} dy.$$

Then completing the square and using Proposition 3.1 gives

$$I = \frac{a}{\sqrt{2\pi}} \int_{-\infty}^{\infty} e^{\frac{-(y+ia\xi)^2}{2}} e^{\frac{-a^2\xi^2}{2}} dy = \frac{a}{\sqrt{2\pi}} \int_{-\infty}^{\infty} e^{\frac{-y^2}{2}} dy \, e^{\frac{-a^2\xi^2}{2}} = a e^{\frac{-a^2\xi^2}{2}}.$$

□

The parameter a^2 from (22) is called the *variance* of the Gaussian random variable. When a^2 is small, things are concentrated near 0, which is the *mean* of the random variable. When a^2 is large, things are spread out. The theorem states that the Fourier transform of a Gaussian is also a Gaussian, but the new variance is the reciprocal of the original variance.

▶ **Exercise 7.8.** Find the Fourier transform of $\frac{\sin(x)}{x}$.

▶ **Exercise 7.9.** Find the Fourier transform of the function that equals 1 on the interval (a, b) and otherwise equals 0.

▶ **Exercise 7.10.** Graph the function $e^{\frac{-x^2}{2a^2}}$ for $a = \frac{1}{3}$, $a = 1$, and $a = 3$.

Let us assume that f is an infinitely differentiable function on **R** and that f decreases rapidly at ∞. For example, f could be any polynomial times the Gaussian $e^{-\frac{x^2}{2}}$. The basic theorems of analysis then allow us to make the following statements rigorous.

1) The Fourier inversion formula holds:

$$(23) \qquad\qquad f(x) = \frac{1}{\sqrt{2\pi}} \int_{0}^{\infty} e^{ix\xi} \hat{f}(\xi) d\xi.$$

2) For each positive integer n it is valid to differentiate (23) n times to obtain

$$(24) \qquad f^{(n)}(x) = (\frac{d}{dx})^n f(x) = \frac{1}{\sqrt{2\pi}} \int_{0}^{\infty} e^{ix\xi} \hat{f}(\xi)(i\xi)^n d\xi.$$

Let us rewrite (24) more abstractly. Let D denote the differentiation operator $\frac{d}{dx}$, and let M denote multiplication by $i\xi$. Let $\mathcal{F}$ denote the operation of taking Fourier transforms. We obtain $D^n f = \mathcal{F}^{-1} M^n \mathcal{F}(f)$. In other words,

$$D^n = \mathcal{F}^{-1} M^n \mathcal{F}.$$

Let α be a positive real number. We define the derivative of order α by

$$(25) \qquad\qquad D^\alpha = \mathcal{F}^{-1} M^\alpha \mathcal{F}.$$

It follows that

$$D^{\alpha+\beta} = D^\alpha D^\beta.$$

The abstract formula (25) gives a definition of a fractional derivative operation! It is difficult to write down an explicit expression for a fractional derivative, and hence we are content with (25). More generally, given a function h, we can try to define $h(D)$ by the formula

$$h(D) = \mathcal{F}^{-1}h(M)\mathcal{F}.$$

4. The Gamma function

The Gamma function extends the factorial function to (most) complex numbers. The reader might pause and start wondering what we could possibly mean by something such as $\frac{3}{2}!$ or $\pi!$, much less $z!$ for $z \in \mathbf{C}$.

Definition 4.1 (The Gamma function as an integral). For $\mathrm{Re}(z) > 0$, we define $\Gamma(z)$ by the formula

(26) $$\Gamma(z) = \int_0^\infty e^{-t}t^{z-1}dt.$$

Let us discuss some technicalities concerning this definition. The integral is *improper* for two reasons. First the upper limit is infinite; no problems result because the decay of e^{-t} at infinity compensates for the growth of t^{z-1} there. Second, when $z - 1$ is negative, there is a singularity at 0. Again no problems result because for $\alpha > -1$, the following limit exists:

$$\lim_{\epsilon \to 0} \int_\epsilon^1 x^\alpha dx.$$

The issue of z being complex is not a problem. Put $z - 1 = \zeta = u + iv$. What do we mean by t^ζ? By definition, since $t > 0$,

$$t^\zeta = e^{\zeta \log(t)} = t^u e^{iv\log(t)}$$

and hence $|t^\zeta| = t^u$. Therefore the imaginary part of ζ does not impact the convergence of the integral.

Note that $\Gamma(1) = \int_0^\infty e^{-t}dt = 1$. We also note, using integration by parts, that if $\mathrm{Re}(z) > 0$, then $\Gamma(z + 1) = z\Gamma(z)$. By combining these facts with induction, we see for $n \in \mathbf{N}$ that $\Gamma(n) = (n - 1)!$. It follows again by induction on n that

(27) $$\Gamma(z + n) = (z + n - 1)(z + n - 2)...(z + 1)(z)\Gamma(z)$$

whenever $\mathrm{Re}(z) > 0$. The functional equation (27) can be used to extend the definition of the Γ function. For example, if $-1 < \mathrm{Re}(z) < 0$, then $\Gamma(z + 1)$ is defined. We can therefore define $\Gamma(z)$ by the formula

$$\Gamma(z) = \frac{\Gamma(z + 1)}{z}.$$

Formula (27) also enables us to define $\Gamma(z)$ whenever the real part of z is not a nonnegative integer. For example we use

$$\Gamma(z) = \frac{\Gamma(z + n)}{(z + n - 1)(z + n - 2)...(z + 1)(z)}$$

to define $\Gamma(z)$ whenever $\mathrm{Re}(z) > -n$.

Remark 4.1. The Γ-function satisfies the following remarkable identity and hence can be defined at all complex numbers except for the negative integers and zero (see [**22**]):

$$(28) \qquad\qquad \Gamma(z) = \lim_{n\to\infty} \frac{n!n^z}{z(z+1)...(z+n)}.$$

The Γ-function arises throughout mathematics, statistics, physics, and engineering. One explanation is simply that it generalizes the factorial. Another explanation is its close connection with the Gaussian. A simple example motivates our next result.

We will show below that $\Gamma(\frac{1}{2}) = \sqrt{\pi}$. Pretending that we know this result, consider the Gaussian integral

$$(29) \qquad\qquad \int_{-\infty}^{\infty} e^{\frac{-x^2}{2}}\, dx.$$

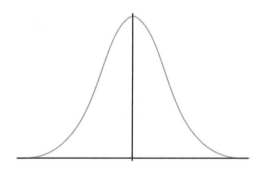

Figure 7.3. The bell curve, or Gaussian.

The next definition is not as general as it could be, but it is adequate for our present purposes.

Definition 4.2. A probability density function is a nonnegative continuous function $f : \mathbf{R} \to \mathbf{R}$ such that $\int_{-\infty}^{\infty} f(t)dt = 1$.

A probability density function defines a *random variable X* by the rule that the probability that X takes a value between a and b equals $\int_a^b f(t)dt$. In order to make $e^{-\frac{x^2}{2}}$ into a probability density function, we must divide by a constant to ensure that the integral over $\mathbf{R}$ equals 1. Thus we must evaluate the integral (29). We noted earlier, in a similar integral without the factor $\frac{1}{2}$, that the evaluation cannot be done by the techniques of one-variable calculus; there is no elementary function whose derivative is $e^{\frac{-x^2}{2}}$. Consider the following approach:

$$(30) \qquad \int_{-\infty}^{\infty} e^{\frac{-x^2}{2}}\, dx = 2\int_0^{\infty} e^{\frac{-x^2}{2}}\, dx = 2\int_0^{\infty} \frac{\sqrt{2}}{2}e^{-t}\frac{dt}{\sqrt{t}} = \sqrt{2}\Gamma(\frac{1}{2}) = \sqrt{2\pi}.$$

In (30) we used $\Gamma(\frac{1}{2}) = \sqrt{\pi}$; we verify this value by a method due to Euler. The technique of multiplying the integrals together generalizes the standard method used in the proof of Proposition 3.1. The far right-hand side of (31) is called the Euler Beta function.

Theorem 4.1. *For* $\mathrm{Re}(a) > 0$ *and* $\mathrm{Re}(b) > 0$ *the following identity holds:*

$$(31) \qquad \frac{\Gamma(a)\Gamma(b)}{\Gamma(a+b)} = \int_0^1 u^{a-1}(1-u)^{b-1}du = \mathcal{B}(a,b).$$

Proof.

$$\Gamma(a)\Gamma(b) = \int_0^\infty e^{-s}s^{a-1}ds \int_0^\infty e^{-t}t^{b-1}dt = \int_0^\infty \int_0^\infty e^{-s+t}s^{a-1}t^{b-1}dsdt.$$

Replace s with x^2 and t with y^2 to obtain

$$\Gamma(a)\Gamma(b) = 4\int_0^\infty \int_0^\infty e^{-(x^2+y^2)}x^{2a-1}y^{2b-1}dxdy.$$

Next use polar coordinates to obtain

$$\Gamma(a)\Gamma(b) = 4\int_0^\infty e^{-r^2}r^{2a+2b-1}\int_0^{2\pi}\cos^{2a-1}(\theta)\sin^{2b-1}(\theta)d\theta.$$

One more change of variables in each integral yields

$$\Gamma(a)\Gamma(b) = \int_0^\infty e^{-t}t^{a+b-1}dt \int_0^1 u^{a-1}(1-u)^{b-1}du = \Gamma(a+b)\int_0^1 u^{a-1}(1-u)^{b-1}du,$$

as desired. $\qquad\square$

Theorem 4.2. *For* $0 < \mathrm{Re}(z) < 1$ *we have*

$$\Gamma(\zeta)\Gamma(1-\zeta) = \frac{\pi}{\sin(\pi\zeta)}.$$

Proof. By Theorem 4.1 it suffices to evaluate $\mathcal{B}(\zeta, 1-\zeta)$. To do so, we first change variables and then use contour integration. We must find

$$\int_0^1 (\frac{u}{1-u})^{\zeta-1}\frac{du}{u}.$$

Putting $x = \frac{u}{1-u}$, we obtain

$$(32) \qquad \int_0^\infty \frac{x^{\zeta-1}}{1+x}dx.$$

We evaluate (32) using the keyhole $\gamma_{\epsilon,R}$ shown in Figure 7.4. By the residue theorem the value of the integral $I_{\epsilon,R}$ is

$$I_{\epsilon,R} = 2\pi i \mathrm{Res}(-1) = 2\pi i(-1)^{\zeta-1} = 2\pi i e^{i\pi(\zeta-1)}.$$

The integrals along the circular arcs tend to 0 as $\epsilon \to 0$ and $R \to \infty$. The integral along the top half of the real axis tends to what we want, namely I; the integral along the bottom half tends to $-e^{2\pi i(\zeta-1)}I$. Putting this information together gives

$$(33) \qquad I(1 - e^{2\pi i(\zeta-1)}) = 2\pi i e^{i\pi(\zeta-1)}.$$

Solving (33) for I and then using the definition of $\sin(\pi\zeta)$ finishes the proof. $\qquad\square$

Corollary 4.1. $\Gamma(\frac{1}{2}) = \sqrt{\pi}$.

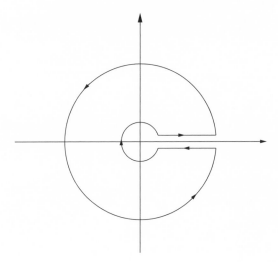

Figure 7.4. Contour for integrals involving branch cuts.

We next combine our knowledge of the Gamma function with our techniques of contour integration to sketch the computation of two important integrals. Choose a real number m with $0 < m < 1$. Consider the real integrals:

$$I_c = \int_0^\infty \frac{\cos(x)}{x^{1-m}}dx,$$

$$I_s = \int_0^\infty \frac{\sin(x)}{x^{1-m}}dx.$$

It is natural to evaluate them simultaneously by finding

(34) $$\int_0^\infty \frac{e^{ix}}{x^{1-m}}dx.$$

To compute (34), we consider a contour γ consisting of the interval (ϵ, R) on the real line, one quarter of a circle of radius R, the part of the imaginary axis from iR to $i\epsilon$, and the quarter circle of radius ϵ. Choose a branch cut avoiding the first quadrant to make z^{m-1} complex analytic there. The integral around γ is then 0 by Cauchy's theorem. As in many of our examples, the integrals on the circular arcs tend to 0 in the limit. The integral along the imaginary axis tends to

$$\int_0^\infty e^{-t}t^{m-1}(i)^m dt,$$

namely $(i)^m\Gamma(m)$. It follows that I_c is the real part of $(i)^m\Gamma(m)$ and I_s is the imaginary part of $(i)^m\Gamma(m)$. Note that $i^m = e^{\frac{im\pi}{2}}$. Hence

$$I_c = \Gamma(m)\cos(\frac{m\pi}{2}),$$

$$I_s = \Gamma(m)\sin(\frac{m\pi}{2}).$$

▶ **Exercise 7.11.** Fill in the details of the above evaluations of I_c and I_s.

Remark 4.2. Often in probability and in combinatorics one needs to know approximately how large $n!$ is for large n. The famous Stirling's formula provides such information. Its proof is beyond the scope of this book (see [**22**]) but we state the result here as an application of the Γ-function:

$$\lim_{n\to\infty} \frac{n!}{n^{n+\frac{1}{2}}e^{-n}} = \sqrt{2\pi}.$$

In other words, $n! = \Gamma(n+1)$ is roughly equal to $\sqrt{2\pi n}(\frac{n}{e})^n$ for large n.

▶ **Exercise 7.12.** Verify that the integrals along the circular arcs in the proof of Theorem 4.2 tend to 0.

▶ **Exercise 7.13.** For positive constants a, λ, the gamma density $g = g_{a,\lambda}$ is defined for $x > 0$ by

$$g(x) = g_{a,\lambda}(x) = c(a,\lambda)x^{a-1}e^{-\lambda x}.$$

Determine the value $c(a,\lambda)$ that makes g into a probability density function; in other words we require $\int_0^\infty g(x)dx = 1$.

▶ **Exercise 7.14.** Determine the Fourier transform of the gamma density from the previous exercise.

▶ **Exercise 7.15.** We can use Stirling's formula to give an approximate formula for the following natural question. For large n, flip a fair coin $2n$ times. Determine approximately the probability of getting exactly n heads.

Chapter 8

Additional Topics

We close this book with short discussions of several additional appealing topics. First we fill in a basic point in analysis. Next we prove the fundamental theorem of algebra: each nonconstant polynomial with complex coefficients has a root. We give three proofs; the first proof could have appeared much earlier in the book, whereas the other two rely on the Cauchy theory. We continue to develop these ideas by introducing winding numbers and using them to give a formula for the number of zeroes and poles of a function inside (the image of) a simple closed curve. These ideas lead to Rouche's theorem, which tells us how to locate (approximately) the zeroes of a polynomial or complex analytic function.

Then we switch gears and briefly discuss several independent topics: Pythagorean triples, elementary mappings, and functions of several complex variables. The section on Pythagorean triples includes an application to the integration of rational functions of cosine and sine. The section on mappings includes a glimpse of non-Euclidean geometry. The section on several variables gives some examples of power series and also resolves one fundamental issue from earlier: we prove that it makes sense to treat z and $\overline{z}$ as independent variables.

1. The minimum-maximum theorem

We first recall some basic analysis. We could have included this material much earlier in the book, but by now it is indispensable. A subset K of $\mathbf{C}$ is closed if and only if its complement is open. By the definition of limit it follows that K is closed if and only if the following holds: whenever $\{z_n\}$ is a sequence in K and $\{z_n\}$ converges to some ζ in $\mathbf{C}$, then $\zeta \in K$. Let $\{z_n\}$ be a sequence and let $k \to n_k$ be an increasing function. We write $\{z_{n_k}\}$ for the subsequence of $\{z_n\}$ whose k-th term is z_{n_k}. All treatments of complex analysis rely on the next result. Its proof uses the method of quadrisection of rectangles. See Section 3 of Chapter 6 for a related use of this technique.

Lemma 1.1. *Let K be a closed bounded subset of $\mathbf{C}$. Then every sequence in K has a convergent subsequence.*

Proof. First we assume that K is a closed rectangle. Assume that $\{z_n\}$ is a sequence in K. We quadrisect K into four smaller closed rectangles. There must be at least one of these four rectangles K_1 such that $z_n \in K_1$ for infinitely many n. Choose $z_{n_1} \in K_1$. Now quadrisect K_1 into four smaller closed rectangles. Again there must be at least one of these rectangles, call it K_2, for which $z_n \in K_2$ for infinitely many n. Choose $n_2 > n_1$ and such that $z_{n_2} \in K_2$. By continuing this process, we find a nested sequence of closed rectangles

$$K \supset K_1 \supset K_2 \cdots \supset K_n \cdots .$$

The lengths of the sides of K_{n+1} are half those of K_n, and hence the side lengths of K_n tend to 0 as n tends to infinity. Note that each K_k contains z_n for infinitely many n. At each stage we may therefore choose $z_{n_k} \in K_k$ such that $k \to n_k$ is an increasing function, giving us a subsequence. It follows by the completeness axiom for the real numbers that the intersection of all these rectangles is a single point ζ. Since $z_{n_k} \in K_k$, it follows that $\{z_{n_k}\}$ converges to ζ.

Now suppose that K is an arbitrary closed and bounded subset of $\mathbf{C}$. Let M be a closed rectangle containing K. Let $\{z_n\}$ be a sequence in K, but regard it as a sequence in M. Then there is a convergent subsequence $\{z_{n_k}\}$ converging to some $\zeta \in M$. Since each $z_{n_k} \in K$ and K is closed, $\zeta \in K$ as well. $\qquad\Box$

Theorem 1.1. *Let K be a closed and bounded subset of $\mathbf{C}$. Assume $f : K \to \mathbf{R}$ is continuous. Then f is bounded. Furthermore there are w and ζ in K such that*

$$(1) \qquad\qquad\qquad f(w) \leq f(z) \leq f(\zeta)$$

for all $z \in K$.

Proof. Assume that f is not bounded. Then we can find a sequence $\{z_n\}$ in K such that $|f(z_n)| > n$. Since K is bounded, there is a subsequence $\{z_{n_k}\}$ converging to some L. Since K is closed, $L \in K$. Now we have $\lim z_{n_k} = L$ but $\lim f(z_{n_k})$ does not exist. Hence f cannot be continuous.

Therefore f continuous on K implies that f is bounded. Let $\alpha = \inf \{f(z) : z \in K\}$ and let $\beta = \sup \{f(z) : z \in K\}$. We can find sequences $\{w_n\}$ and $\{\zeta_n\}$ such that $f(w_n)$ converges to α and $f(\zeta_n)$ converges to β. Since K is bounded, both $\{w_n\}$ and $\{\zeta_n\}$ are themselves bounded sequences. Hence they have convergent subsequences, with limits w and ζ. Since $w_{n_k} \to w$ and f is continuous, we have $\alpha = f(w)$. Similarly $\beta = f(\zeta)$. Thus (1) holds. $\qquad\Box$

2. The fundamental theorem of algebra

In this section we prove that every nonconstant polynomial has a complex root. The proof does not rely on the Cauchy theory, and hence we could have included it earlier. We suggest a second proof in the exercises; the second proof appears in most texts and relies on the Cauchy theory. It passes through Liouville's theorem, which follows from the Cauchy integral formula. We give a third proof in Section 3.

The fundamental theorem of algebra is a bit of a misnomer. First of all, despite its elegance and simplicity, not all algebraists regard it as *the* fundamental result in the subject. Second of all, one cannot prove it using algebra alone. All proofs of it must rely on the completeness axiom for the real number system. Our proof uses Theorem 1.1, guaranteeing that continuous real-valued functions achieve their minima and maxima on closed bounded sets.

Theorem 2.1 (Fundamental theorem of algebra). *Let p be a nonconstant polynomial. Then p has a complex root.*

Proof. Seeking a contradiction, we suppose that $p(z)$ is never 0. The first step is to find a point w such that $0 < |p(w)| \leq |p(z)|$ for all $z \in \mathbf{C}$. First we consider z with $|z|$ large. Note that $\lim_{z \to \infty}(p(z)) = \infty$. By the definitions of limits involving infinity, there is a positive real number R such that $|p(z)| > |p(0)|$ for $|z| \geq R$. Next we consider z inside this disk. The function $z \to |p(z)|$ is continuous. By Theorem 1.1 the minimum value of $z \to |p(z)|$ on the closed disk $|z| \leq R$ is achieved at some point w. By assumption $p(w) \neq 0$. Thus $0 < |p(w)| \leq |p(z)|$ for $|z| \leq R$, including $z = 0$. But for $|z| \geq R$ we have $|p(w)| \leq |p(0)| < |p(z)|$. Thus $|p(w)| \leq |p(z)|$ for all z in $\mathbf{C}$. We have completed the first step.

We may multiply p by an appropriate nonzero constant and assume therefore, without loss of generality, that $p(w) = 1$. Consider the polynomial function f given by $f(\zeta) = p(w + \zeta)$. By our assumptions on p, the absolute value of f has a minimum of 1 at $\zeta = 0$. The second step is to obtain a contradiction; we do so by finding a point ζ where $|f(\zeta)| < 1$.

There is a positive integer m such that we can write $f(\zeta) = 1 + c\zeta^m + g(\zeta)$, where $c \neq 0$ and all terms in g are of degree larger than m. Thus the limit of $\frac{c\zeta^m + g(\zeta)}{\zeta^m}$ as ζ tends to 0 is the nonzero number c. Hence there is a $\delta > 0$ such that $|\zeta| < \delta$ implies $|g(\zeta)| < \frac{1}{2}|c||\zeta|^m$. Intuitively, for ζ small, the term $c\zeta^m$ dominates $g(\zeta)$. Assuming $0 < |\zeta| < \delta$, we write $\zeta = |\zeta|e^{i\theta}$ and choose θ in order to make $c\zeta^m$ equal to a negative number, say $-\alpha$, with $0 < \alpha < 1$. With this choice of ζ, we have $|f(\zeta)| = 1 - \alpha + \beta$, where $|\beta| < \frac{\alpha}{2}$. The triangle inequality implies

$$(2) \qquad |p(w + \zeta)| = |f(\zeta)| = |1 - \alpha + \beta| \leq 1 - \alpha + |\beta| < 1 - \alpha + \frac{\alpha}{2} < 1.$$

We have found a z for which $|p(z)| < 1$, a contradiction. It must therefore be true that p is 0 somewhere. $\square$

The intuitive ideas behind this proof are simpler than the details. First of all, $|p(z)|$ is large when $|z|$ is large. Hence we can find an R such that the minimum of $|p|$ on all of $\mathbf{C}$ happens on the disk $|z| \leq R$. By Theorem 1.1 this minimum is guaranteed to exist on a closed disk. Assume it occurs at w. If we assume that this minimum is not 0, then after dividing by a constant we may assume that $p(w) = 1$. We obtain a contradiction by finding a direction ζ such that moving from w in the ζ direction *decreases* $|p|$. To find this direction, we need only solve the simpler polynomial equation $c\zeta^m = -\alpha$.

We now turn to some simple consequences of the fundamental theorem of algebra. Once we know that a polynomial has a root at a, we can divide by $(z - a)$ and the quotient is also a polynomial. This fact is much more elementary than is the fundamental theorem of algebra. In fact, in Exercise 1.35 you were asked to prove a stronger statement in the context of real polynomials. The proof suggested there relies on the division algorithm for real polynomials and would also work here. We give next a direct different proof that the quotient is a polynomial. After doing so, we combine the result with the existence of one root to show that a polynomial of degree d has d roots, as long as we take multiplicity into account.

Proposition 2.1. *Let p be a polynomial of degree $d \geq 1$. Suppose $p(a) = 0$. Then there is a polynomial q of degree $d - 1$ such that $p(z) = (z - a)q(z)$.*

Proof. First we consider the special case where $p(z) = z^n - a^n$. In this case we can divide explicitly to get

$$(3) \qquad \frac{z^n - a^n}{z - a} = z^{n-1} + z^{n-2}a + z^{n-3}a^2 + \ldots + a^{n-1} = h_{n-1}(z),$$

where h_{n-1} is of degree $n - 1$. Notice that we have not indicated its dependence on a. Let now p be an arbitrary polynomial of degree d. Since the constant term drops out when we subtract, we can write

$$(4) \qquad p(z) = p(z) - p(a) = \sum_{n=0}^{d} c_n(z^n - a^n) = \sum_{n=1}^{d} c_n(z^n - a^n).$$

By combining (3) and (4), we find an explicit polynomial q for which

$$(5) \qquad p(z) = \sum_{n=1}^{d} c_n(z - a)h_{n-1}(z) = (z - a)\sum_{n=1}^{d} c_n h_{n-1}(z) = (z - a)q(z).$$

$\square$

Corollary 2.1. *Let p be a polynomial of degree d. With multiplicity counted, p has d roots in $\mathbf{C}$.*

Proof. By the fundamental theorem of algebra, p has at least one root a. By Proposition 2.1, we can write $p(z) = (z - a)q(z)$, where the degree of q is $d - 1$. The result therefore follows by induction on d. $\square$

▶ **Exercise 8.1.** Suppose p and q are polynomials of degree at most d and they agree at $d + 1$ values. Show that $p = q$.

▶ **Exercise 8.2.** Put $p(z) = z^3 + 2z^2 + 2z - 5$. Note that $p(1) = 0$. Find the quotient $\frac{p(z)}{z-1}$ both by long division and by the method from Proposition 2.1.

An alternative proof of the fundamental theorem of algebra is based on Liouville's theorem, which we derive from the Cauchy integral formula. We use the standard term *entire analytic function* for a function which is complex analytic on all of $\mathbf{C}$.

Theorem 2.2 (Liouville's theorem). *A bounded entire analytic function must be a constant.*

Proof. Assume that $|f(z)| \leq M$ for all z. Fix a point z. Start with the Cauchy integral formula for $f(z)$, where the integral is taken over a circle of radius R about z. As in Corollary 4.1 from Chapter 6, differentiate once to find a formula for $f'(z)$:

$$f'(z) = \frac{1}{2\pi i} \int_{|\zeta - z| = R} \frac{f(\zeta) d\zeta}{(\zeta - z)^2}.$$

Estimate $|f'(z)|$ using the ML-inequality and the bound for $|f|$ (as in the Cauchy estimates) to get $|f'(z)| \leq \frac{M}{R}$. Let R tend to infinity and conclude that $f'(z) = 0$. Since z is arbitrary, f is a constant. $\qquad\square$

▶ **Exercise 8.3.** Derive the fundamental theorem of algebra from Liouville's theorem. Hint: If p were never zero, then $\frac{1}{p}$ would be an entire analytic function.

▶ **Exercise 8.4.** Suppose f is an entire analytic function and $|f(z)| \geq 1$. Prove that f is a constant.

▶ **Exercise 8.5.** Give an example of a nonconstant entire analytic function f such that $f(z) \neq 0$ for all z.

Remark 2.1. The conclusion of Liouville's theorem holds under much weaker hypotheses than boundedness. Picard's theorem states that an entire analytic function that misses two values must be a constant. See [**1**].

▶ **Exercise 8.6.** Let f be an entire analytic function satisfying the inequality

$$|f(z)| \leq c|z|^m$$

for some positive integer m, some positive constant c, and for $|z|$ sufficiently large. What can you conclude about f? Hint: Follow the proof of Liouville's theorem. Use Corollary 4.2 from Chapter 6, but differentiate $m + 1$ times.

▶ **Exercise 8.7.** Prove that a bounded entire harmonic function is a constant.

3. Winding numbers, zeroes, and poles

In this section we develop a method for counting the number of zeroes and poles of a function in a region. A third proof of the fundamental theorem of algebra will result. Even better we will prove Rouche's theorem, which gives a good way to locate approximately where the zeroes of a polynomial actually are.

The functions to be considered will be complex analytic, on and inside a simple closed curve γ, with the possible exception of a finite number of points in the interior of the curve, at which the functions may have poles. To get started, consider the most naive example. Suppose $f(z) = z^n$, where $n \in \mathbf{Z}$. If $n > 0$, then f has a zero of order n at 0. If $n < 0$, then f has a pole of order n there. Let γ be the unit circle, traversed once with positive orientation. We can determine the exponent n by doing a line integral:

$$(6) \qquad \frac{1}{2\pi i} \int_\gamma \frac{f'(z)}{f(z)} dz = \frac{1}{2\pi i} \int_\gamma \frac{n}{z} dz = n.$$

Formula (6) applies even when $n = 0$. For $n > 0$, the value of the integral on the left-hand side of (6) equals the number of zeroes of f inside with multiplicity counted. For $n = 0$, the same is true. For $n < 0$, we get the same result if we regard a pole of order n as a zero of order $-n$. When f has a pole of order m at c, we say that f has m poles, with multiplicity counted, there.

The same idea works in more generality. First consider a function of the following form:

$$f(z) = \frac{\prod_{j=1}^{m}(z - a_j)}{\prod_{k=1}^{n}(z - b_k)}.$$

We allow repetitions among the a_j or among the b_k, but we may assume that $a_j \neq b_k$ for all j, k, because we can cancel common factors. We compute $\frac{f'}{f}$, either by the rules of calculus or by logarithmic differentiation, to get

(7)
$$\frac{f'(z)}{f(z)} = \sum_{j=1}^{m} \frac{1}{z - a_j} - \sum_{k=1}^{n} \frac{1}{z - b_k}.$$

Let us integrate (7) around a curve (satisfying the usual hypotheses) enclosing all the a_j and b_k. Each a_j contributes a factor of $2\pi i$ and each b_k contributes a factor of $-2\pi i$. Thus we recover the number of zeroes minus the number of poles (both with multiplicity counted) by integrating (7). We have the following general result:

Theorem 3.1. *Let γ be a simple closed positively oriented curve surrounding the region Ω. Suppose that h is complex analytic on and inside γ, except for a finite set of points $b_j \in \Omega$ at which h has poles. Assume also that $h \neq 0$ on γ. Let $Z(h)$ be the number of zeroes of h inside γ with multiplicity counted, and let $P(h)$ be the number of poles of h inside γ again with multiplicity counted. Then*

(8)
$$Z(h) - P(h) = \frac{1}{2\pi i} \int_{\gamma} \frac{h'(z)}{h(z)} dz.$$

Proof. Suppose first that H is complex analytic on and inside γ and not zero on γ. We first verify that (8) holds for H, where $P(H) = 0$. By Corollary 4.5 of Chapter 6 of the Cauchy integral formula, H has at most finitely many zeroes $a_1, ..., a_{N_1}$ inside. Then by Theorem 4.4 of Chapter 6 we have

(9)
$$H(z) = \prod_{j=1}^{N_1} (z - a_j)\, u(z),$$

where $u(z) \neq 0$. Taking derivatives gives

(10)
$$\frac{H'(z)}{H(z)} = \sum_{j=1}^{N_1} \frac{1}{z - a_j} + \frac{u'(z)}{u(z)}.$$

Integrate both sides of (10) around γ. Since $u(z) \neq 0$, the function $\frac{u'}{u}$ is complex analytic. By Cauchy's theorem, $\int_{\gamma} \frac{u'(z)}{u(z)} dz = 0$. Therefore, integrating (10) gives

(11)
$$\frac{1}{2\pi i} \int_{\gamma} \frac{H'(z)}{H(z)} dz = \frac{1}{2\pi i} \sum_{j=1}^{N_1} \int_{\gamma} \frac{1}{z - a_j} dz = N_1 = Z(H).$$

Next suppose that h has poles. We assume these poles are at $c_1, ..., c_k$. If the order of the pole at c_j is m_j, then we list it m times as some of the b_j. In this way assume that the poles are at b_j for $j = 1, ..., N_2$ and $P(h) = N_2$. Define H by

$$H(z) = \prod_{j=1}^{N_2} (z - b_j)h(z).$$

Then H has the same zeroes as h but no poles. Thus $Z(h) = Z(H)$ and $P(H) = 0$. We may apply (11) to H. Taking derivatives yields

$$(12) \qquad \frac{H'(z)}{H(z)} = \sum_{j=1}^{N_2} \frac{1}{z - b_j} + \frac{h'(z)}{h(z)}.$$

Now integrating (12) and using (11) shows that

$$(13) \qquad Z(h) = Z(H) = N_2 + \frac{1}{2\pi i} \int_\gamma \frac{h'(z)}{h(z)} dz = P(h) + \frac{1}{2\pi i} \int_\gamma \frac{h'(z)}{h(z)} dz.$$

Hence formula (8) holds for h. $\qquad\qquad\qquad\qquad\qquad\qquad\qquad\qquad\qquad\Box$

We next discuss winding numbers and the *argument principle*. These ideas make the previous result geometric. The expression $\frac{h'}{h}$ is the logarithmic derivative of h. In a sense that we will make precise, the integral in (8) is computing the winding number of the image curve $h(\gamma)$ about the origin. Each zero that occurs forces the image curve to wind around zero in a counterclockwise fashion, and each pole forces the image curve to wind around zero in a clockwise fashion. In Figure 8.1 we illustrate this principle with a simple example. The word *argument* refers to the polar angle. The argument principle says that the line integral (8) measures the total change in the argument as we traverse γ.

Example 3.1. Put $f(z) = z(2z - 1)(3z - 2)$ and let γ be the unit circle. Then f has three zeroes inside γ. Furthermore, as indicated in Figure 8.1, the image of γ under f winds around zero a total of three times.

The following crucial result uses integrals to make the notion of winding number precise. We will prove it under the weaker assumption that the curve γ is smooth. See [**1, 19**] for the general case.

Theorem 3.2. *Let γ be a continuous closed curve (not assumed positively oriented or simple) in $\mathbf{C}$, and suppose p does not lie on (the image of) γ. The value of the following integral is an integer, called the winding number of γ about p:*

$$\mathbf{n}(\gamma, p) = \frac{1}{2\pi i} \int_\gamma \frac{dz}{z - p}.$$

Proof. As mentioned above, we prove the result only when γ is a smooth curve. Suppose $\gamma : [0, 1] \to \mathbf{C}$. The proof amounts to using the idea of the logarithm without actually using logarithms. We will show that $e^{2\pi i \mathbf{n}(\gamma, p)} = 1$. Since γ is smooth, the winding number is given by

$$\mathbf{n}(\gamma, p) = \frac{1}{2\pi i} \int_0^1 \frac{\gamma'(u)}{\gamma(u) - p} du.$$

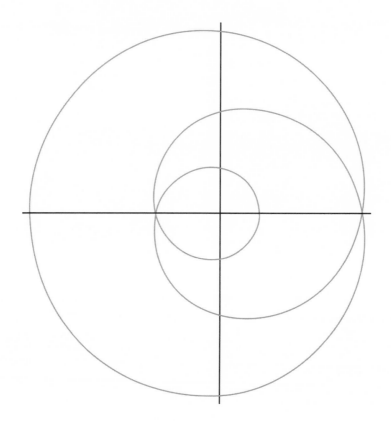

Figure 8.1. Winding number from Example 3.1.

Consider the function ϕ defined by

$$\phi(t) = \frac{1}{2\pi i} \int_0^t \frac{\gamma'(u)}{\gamma(u) - p} du.$$

By the fundamental theorem of calculus, ϕ is differentiable and

$$(14) \qquad\qquad \phi'(t) = \frac{1}{2\pi i} \frac{\gamma'(t)}{\gamma(t) - p}.$$

Thinking heuristically, we expect that $2\pi i \phi(t)$ should satisfy

$$2\pi i \phi(t) = \log(\gamma(t) - p) - \log(\gamma(0) - p) = \log\left(\frac{\gamma(t) - p}{\gamma(0) - p}\right),$$

and hence we should have

$$(15) \qquad\qquad e^{2\pi i \phi(t)} = \frac{\gamma(t) - p}{\gamma(0) - p}.$$

Recall that γ is closed and hence $\gamma(1) = \gamma(0)$. If (15) holds, then we obtain

$$e^{2\pi i \phi(1)} = \frac{\gamma(1) - p}{\gamma(0) - p} = 1,$$

and hence $\mathbf{n}(\gamma, p) = \phi(1)$ is an integer.

We establish (15) by differentiation. Consider

$$g(t) = e^{-2\pi i \phi(t)} \frac{\gamma(t) - p}{\gamma(0) - p}.$$

Then $g(0) = 1$. If we show g is a constant, then we are done. But, differentiating and using (14), we get

$$g'(t) = e^{-2\pi i \phi(t)} \left((-2\pi i \phi'(t)) (\frac{\gamma(t) - p}{\gamma(0) - p}) + \frac{\gamma'(t)}{\gamma(0) - p} \right) = 0.$$

$\square$

The reader should contemplate the several ways in which this proof exhibits the same spirit as the proof we gave in Chapter 3 of the identity $\log(ab) = \log(a) + \log(b)$ for positive a and b.

The notion of winding number extends to finite sums of closed curves. If $\gamma = \gamma_1 + \gamma_2$ and ω is a differential 1-form, then we put $\int_\gamma \omega = \int_{\gamma_1} \omega + \int_{\gamma_2} \omega$. When $\omega = \frac{1}{2\pi i} \frac{dz}{z-p}$, we obtain the winding number $\mathbf{n}(\gamma, p)$ of γ about p.

We can now make precise the notion that an open connected set have no holes. Somehow the vague concept of hole, the winding number, and the multi-valued aspects of the logarithm function all describe the same phenomenon.

Definition 3.1. An open connected subset Ω is called *simply connected* if $\mathbf{n}(\gamma, p) = 0$ for all sums γ of closed curves in Ω and all points p not in Ω.

The idea of winding number enables us to interpret the formula for zeroes and poles. Suppose that h is complex analytic on and inside a simple closed curve γ, except possibly for finitely many poles inside. Then $Z(h) - P(h)$ equals the number of times $h(\gamma)$ winds around the origin. To verify this fact, we change variables in the integral from (8), setting $w = h(z)$. Then $\frac{dw}{w} = \frac{h'(z)}{h(z)} dz$. Thus the integral in (8) is the same as $\frac{1}{2\pi i} \int_{h(\gamma)} \frac{dw}{w} = \mathbf{n}(h(\gamma), 0)$; in other words, the integral represents how many times $h(\gamma)$ winds around the origin.

These ideas yield a third proof of the fundamental theorem of algebra. The intuition behind this particular proof is beautiful. For $n \geq 1$, consider the polynomial $z^n + a(z)$, where the degree of $a(z)$ is less than n. For large z the polynomial $z^n + a(z)$ behaves much like z^n. Since we can recover the values of a complex analytic function inside of a circle by knowing its values on the circle, it is at least plausible that the polynomial $a(z)$ serves as a minor perturbation of z^n. Hence, while a impacts the location of the zeroes, it does not impact their existence. Let us make this argument precise. In the next theorem we think of g as a small perturbation of f.

Theorem 3.3 (Rouche's theorem). *Assume that f and g are complex analytic on and inside a smooth simple closed curve γ and that $|f(z) - g(z)| < |f(z)|$ on γ. Then f and g have the same number of zeroes inside γ.*

Proof. By using (8), we see that f and g have the same number of zeroes inside γ if

(16)
$$\int_\gamma \frac{f'(z)}{f(z)}\,dz = \int_\gamma \frac{g'(z)}{g(z)}\,dz.$$

Put $h = \frac{g}{f}$. By the quotient rule for derivatives we have

(17)
$$\frac{h'}{h} = \frac{fg' - gf'}{fg} = \frac{g'}{g} - \frac{f'}{f}.$$

The equality in (16) yields the following condition. The complex analytic functions f and g have the same number of zeroes inside γ if the function h has the same number of zeroes as poles. The inequality $|f(z) - g(z)| < |f(z)|$ on γ guarantees this conclusion, for the following reason. This inequality is equivalent to $|1 - h(z)| < 1$ for $z \in \gamma$. Hence the image curve is contained in a disk of radius one about 1. In particular the image curve does not wind around zero. Therefore $\int_\gamma \frac{h'(z)}{h(z)}\,dz = 0$, and hence $Z(f) = Z(g)$. □

Remark 3.1. The inequality $|f - g| < |f|$ on γ can be weakened to $|f - g| < |f| + |g|$, but we do not need this refinement.

Corollary 3.1 (The fundamental theorem of algebra). *For $n \geq 1$ a polynomial of degree n has (with multiplicity counted) n complex roots.*

Proof. We may assume that $p(z) = z^n + a(z)$ where the degree of a is at most $n - 1$. Then $\lim_{z \to \infty} \frac{a(z)}{z^n} = 0$ and we have, for large R, $|a(z)| < |z|^n$ for $|z| = R$. By Rouche's theorem, the number of zeroes of p and of z^n inside $|z| = R$ are equal. □

▶ **Exercise 8.8.** How many roots of $z^3 - 20z + 5$ lie inside $|z| = 5$? How many lie in the annulus $1 < |z| < 4$?

▶ **Exercise 8.9.** How many roots of $z^5 - 20z + 2$ lie inside $|z| < 3$? How many lie inside $|z| < 1$?

▶ **Exercise 8.10.** How many roots of $z^5 + 2z^3 + 7z + 1$ lie inside $|z| = 2$? How many lie inside $|z| < 1$? How many roots of this polynomial are real? (Suggestion: Look at the derivative.)

▶ **Exercise 8.11.** Suppose g is complex analytic on and inside γ and h is as above. Evaluate the integral

$$\frac{1}{2\pi i} \int_\gamma \frac{h'(z)}{h(z)} g(z)\,dz$$

in terms of the zeroes and poles of h inside γ. (Comment: This idea arises in one proof of the Weierstrass preparation theorem from several complex variables. See page 21 of [**6**].)

▶ **Exercise 8.12.** Find all polynomials p such that $|p(z)| = 1$ on $|z| = 1$.

▶ **Exercise 8.13.** Suppose a polynomial p has real coefficients and is of odd degree. What can you say about the number of its real roots?

▶ **Exercise 8.14.** Use Rouche's theorem to give a simple criterion for a complex analytic function to have precisely one fixed point in the unit disk. (A fixed point is a solution to $f(z) = z$.)

▶ **Exercise 8.15.** Show that the annulus defined by $r_1 < |z| < r_2$ is not simply connected.

4. Pythagorean triples

Recall that a triple (a, b, c) of nonnegative integers is called a Pythagorean triple if $a^2 + b^2 = c^2$. Equivalently we have

$$(18) \qquad (\frac{a}{c})^2 + (\frac{b}{c})^2 = 1.$$

Based on (18), we start by thinking about the unit circle. Assume (a, b, c) is a Pythagorean triple. Consider the complex number $z = \frac{a+ib}{c}$. By (18), z lies on the unit circle, and both its real and imaginary parts are rational numbers. We call z a rational point on the circle. In order to find Pythagorean triples, we study rational points on the circle.

The starting point involves parametrizing the unit circle in a different fashion from the usual cosine and sine and also distinct from the complex exponential. See (4) of Chapter 3. Consider the line with slope t through the point -1; its equation is $y = t(x+1)$. By geometric considerations, this line intersects the circle in exactly one other point. We can find that point by solving the system of equations

$$(19) \qquad y = t(x + 1),$$

$$(20) \qquad x^2 + y^2 = 1$$

for x and y as functions of t. We obtain $x^2 + t^2(x + 1)^2 = 1$ from which we get $x = -1$ or $x = \frac{1-t^2}{1+t^2}$. We disregard the value $x = -1$ because that value gives us back the point -1. Using $y = t(x + 1)$, we obtain, for $-\infty < t < \infty$,

$$(21) \qquad z(t) = x(t) + iy(t) = \frac{1 - t^2}{1 + t^2} + i\frac{2t}{1 + t^2} = \frac{(1 + it)^2}{|1 + it|^2} = \frac{1 + it}{1 - it}.$$

The formula in (21) parametrizes the unit circle except for the point $z = -1$, which we recover by setting t equal to infinity. The mapping $t \to z(t)$ is called a *birational isomorphism* from the extended real line to the unit circle. If t is rational, then $z(t)$ has rational coordinates, and if $z(t)$ has rational coordinates, then t is rational.

In a moment we will use (21) to find all Pythagorean triples. First we give an application to basic calculus. We do not give a precise definition of elementary function. Roughly speaking, we mean a function that can be expressed as a formula involving polynomials, trigonometric functions, the exponential function, and their inverses.

Proposition 4.1. *Let $R(x, y)$ be a rational function (the quotient of polynomials). Then the indefinite integral*

$$I = \int R(\cos(\theta), \sin(\theta))d\theta$$

is an elementary function.

Proof. To evaluate I, we substitute $t = \tan(\frac{\theta}{2})$. Thus $\theta = 2\tan^{-1}(t)$ and

$$(22) \qquad\qquad d\theta = \frac{2dt}{1+t^2}.$$

The equations in (19) and (20) give the formulas

$$(23) \qquad\qquad \cos(\theta) = \frac{1-t^2}{1+t^2},$$

$$(24) \qquad\qquad \sin(\theta) = \frac{2t}{1+t^2}.$$

The indicated change of variables in (22) involves only rational functions and it results in the integral of a rational function in t times dt. The integral of a rational function can always be done in terms of elementary functions via partial fractions. Substituting back for θ in terms of t finishes the proof of the proposition. $\square$

▶ **Exercise 8.16.** Evaluate $\int \sec(\theta)d\theta$ using Proposition 2.1.

▶ **Exercise 8.17.** Find a rational parametrization for the curve defined by $|z|^4 = \operatorname{Re}(z^2)$. Graph the curve.

We use the mapping defined by (21) to find all Pythagorean triples. In writing the triple (a, b, c), we always assume that a, b, c are positive numbers. We say that a triple is *primitive* if a, b, c have no common integer factors larger than 1.

First suppose $t = \frac{p}{q}$ is rational; then $z(t)$, as defined in (21), lies on the unit circle and has rational coordinates $\frac{q^2-p^2}{p^2+q^2}$ and $\frac{2pq}{p^2+q^2}$. If also $q > p$, that is, $0 < t < 1$, then the triple $(q^2 - p^2, 2pq, q^2 + p^2)$ is a Pythagorean triple. Conversely, suppose that (a, b, c) is a Pythagorean triple. Then we have

$$(25) \qquad\qquad \frac{a}{c} = \frac{1-t^2}{1+t^2},$$

$$(26) \qquad\qquad \frac{b}{c} = \frac{2t}{1+t^2}.$$

By (25), we have $t^2 = \frac{c-a}{a+c}$. Plugging this value into (26) gives $t = \frac{b}{a+c}$. Finally, if we put $p = b$ and $q = a + c$, then the triple $(q^2 - p^2, 2pq, q^2 + p^2)$ becomes (after some simplification) $2(a + c)(a, b, c)$.

We summarize these computations as follows.

Theorem 4.1. *Let $t = \frac{p}{q}$ be rational with $0 < t < 1$. Then $(q^2 - p^2, 2pq, q^2 + p^2)$ is a Pythagorean triple. Let (a, b, c) be a Pythagorean triple. Put $t = \frac{b}{a+c} = \frac{p}{q}$. Then $(q^2 - p^2, 2pq, q^2 + p^2)$ is a multiple of (a, b, c).*

Corollary 4.1. *All Pythagorean triples are of the form*

$$\frac{1}{n}(q^2 - p^2, 2pq, q^2 + p^2)$$

for integers p, q, n, where p and q have no common factors.

We make some additional remarks. Let (a, b, c) be a primitive Pythagorean triple. Then c is odd, and precisely one of a or b is odd. See Exercise 8.18. Assume that a is odd; since $a > 1$, we can write $a = uv$ for odd numbers u, v with $u > v \geq 1$. Put $q = \frac{u+v}{2}$ and $p = \frac{u-v}{2}$. We then obtain the alternative form

$$(27) \qquad (a, b, c) = (uv, \frac{u^2 - v^2}{2}, \frac{u^2 + v^2}{2}).$$

The formula (27) shows that we can always take the integer n from Corollary 4.1 equal to either 1 or 2.

Let us discuss briefly some issues arising because rational numbers need not be in lowest terms. Given a triple (a, b, c) and a positive integer n, of course (na, nb, nc) is also a triple. These two triples determine the same t, because $\frac{nb}{na+nc} = \frac{b}{a+c}$. On the other hand, given a rational number t expressed in lowest terms $\frac{p}{q}$, the resulting triple

$$(a, b, c) = (q^2 - p^2, 2pq, q^2 + p^2)$$

need not be primitive.

One also needs to be a bit careful about the order. For example, for the triple $(3, 4, 5)$ we have $t = \frac{1}{2}$, whereas for the triple $(4, 3, 5)$ we have $t = \frac{1}{3}$. One annoying point in this development is that the expression $(q^2 - p^2, 2pq, q^2 + p^2)$ might be a multiple of a triple; for example the triple $(4, 3, 5)$ appears as $(8, 6, 10)$. If we solve the equations

$$(q^2 - p^2, 2pq, q^2 + p^2) = (4, 3, 5),$$

for p and q, then we get $q = \frac{3}{\sqrt{2}}$ and $p = \frac{1}{\sqrt{2}}$.

We end by listing the triples corresponding to various rational numbers $t = \frac{p}{q}$. The list includes some redundancies in order to illustrate some of the above ideas. For example, note that $(8, 6, 10)$ arises instead of $(4, 3, 5)$.

$$(q, p) = (2, 1) \to (3, 4, 5),$$
$$(q, p) = (3, 1) \to (8, 6, 10),$$
$$(q, p) = (3, 2) \to (5, 12, 13),$$
$$(q, p) = (4, 1) \to (15, 8, 17),$$
$$(q, p) = (4, 2) \to (12, 16, 20),$$
$$(q, p) = (4, 3) \to (7, 24, 25),$$
$$(q, p) = (5, 1) \to (24, 10, 26),$$
$$(q, p) = (5, 2) \to (21, 20, 29),$$
$$(q, p) = (5, 3) \to (16, 30, 34),$$
$$(q, p) = (5, 4) \to (9, 40, 41),$$
$$(q, p) = (n, 1) \to (n^2 - 1, 2n, n^2 + 1),$$
$$(q, p) = (2k + 1, 2) \to (4k^2 + 4k - 3, 8k + 4, 4k^2 + 4k + 5).$$

▶ **Exercise 8.18.** Suppose (a, b, c) is a Pythagorean triple. Show by elementary means that both a and b cannot be odd. (Work modulo 4.) If (a, b, c) is primitive, show that c must be odd and exactly one of a or b is odd.

▶ **Exercise 8.19.** Show that $(1, b, c)$ cannot be a Pythagorean triple.

▶ **Exercise 8.20.** Find all Pythagorean triples of the form $(a, b, b + 1)$.

▶ **Exercise 8.21.** Suppose s, t are rational and consider the corresponding rational points on the circle $z(s) = \frac{1+is}{1-is}$ and $z(t) = \frac{1+it}{1-it}$ Then $z(s)z(t)$ is also a rational point on the unit circle. What does this fact say about Pythagorean triples?

▶ **Exercise 8.22.** Put $z = q + ip$. Show that the triple $(q^2 - p^2, 2pq, q^2 + p^2)$ can be written

$$(\mathrm{Re}(z^2), \mathrm{Im}(z^2), |z|^2).$$

▶ **Exercise 8.23.** Let a, b, c, d be integers. Show that $(a^2 + b^2)(c^2 + d^2)$ is a sum of two squares. (Suggestion: Consider $a + ib$ and $c + id$.)

5. Elementary mappings

This section consists primarily of a few exercises of the following sort. Given domains Ω_1 and Ω_2 in **C** with simple geometry, we seek a complex analytic function $f : \Omega_1 \to \Omega_2$. We want f to be a bijection; it follows that the inverse function is also complex analytic. In this situation we say that Ω_1 and Ω_2 are *conformally equivalent* or *biholomorphically equivalent*.

The subject is well developed and is important in physics and engineering, but it is beyond the scope of this book. We say annoyingly little about it! We do state the Riemann mapping theorem. We describe the conformal mappings of the unit disk, and we mention their connection to non-Euclidean geometry.

▶ **Exercise 8.24.** Fix some nonzero complex number w. Define $f : \mathbf{C} \to \mathbf{C}$ by $f(z) = wz$. Show that f preserves angles, in the following sense: if u and v are vectors based at z and $f(u)$ and $f(v)$ are the vectors based at $f(z)$, show that the angle between u and v equals the angle between $f(u)$ and $f(v)$. What happens if, instead, $f(z) = \overline{z}$?

Assume that f is complex analytic on an open set Ω. For each $z \in \Omega$ and sufficiently small ζ we can write

$$f(z + \zeta) = f(z) + f'(z)\zeta + E(z, \zeta),$$

where the error term $E(z, \zeta)$ is small in the sense that

$$\lim_{\zeta \to 0} \frac{E(z, \zeta)}{\zeta} = 0.$$

Infinitesimally, f is just multiplication by the number $f'(z)$. By the previous exercise, it follows when $f'(z) \neq 0$ that, infinitesimally, f is a conformal mapping. The standard definition of conformality follows:

Definition 5.1. Suppose $f : \Omega_1 \to \Omega_2$. Then f is *conformal* if f is complex analytic and one-to-one.

▶ **Exercise 8.25.** Find a linear fractional transformation that maps the interior of a circle of radius 2 with center at 2 to the exterior of a circle of radius 1 centered at i.

▶ **Exercise 8.26.** Consider the region in the first quadrant bounded by the four hyperbolas $x^2 - y^2 = 1$, $x^2 - y^2 = 2$, $xy = 1$, $xy = 2$. Find a conformal mapping from this region to the interior of a rectangle.

▶ **Exercise 8.27.** Consider the region inside of a circle passing through the origin and otherwise lying in the right half-plane. Find a conformal mapping from it to the upper half-plane.

▶ **Exercise 8.28.** Consider the infinite strip given by $0 < \mathrm{Im}(z) < \pi$. Find a (conformal) mapping that sends this strip to the upper half-plane.

▶ **Exercise 8.29.** Consider the semi-infinite strip given by $-\frac{\pi}{2} < \mathrm{Re}(z) < \frac{\pi}{2}$ and $\mathrm{Im}(z) > 0$. Find the image of this strip under the mapping $w = \sin(z)$.

▶ **Exercise 8.30.** Consider the region Ω given by $x > 0$, $y > 0$, and $xy < 1$. Graph Ω. Find the image of Ω under the mapping given by $w = z^2$. Then find a formula for a conformal mapping f from Ω to the upper half-plane.

▶ **Exercise 8.31.** Let $f(z) = z + \frac{1}{z}$. Find the image of the top half of the unit disk under f.

We next state the famous Riemann mapping theorem. See [**1, 10, 19**] for more details and the proof. A domain in **C** is an open connected set. A domain is called *simply connected* if it has no holes. See Definition 3.1. We note that the region between concentric circles is *not* simply connected and that the interior of a simple closed curve is simply connected. We let **B** denote the open unit disk.

Theorem 5.1 (Riemann mapping theorem). *Let Ω be a simply connected domain in* **C**; *assume that Ω is not all of* **C**. *Then there is a bijective complex analytic mapping $f : \Omega \to \mathbf{B}$.*

The derivative f' of a bijective complex analytic mapping f is not zero; hence f is conformal. In Lemma 1.1 of Chapter 3 we wrote down an explicit conformal mapping from the upper half-plane to the unit disk. See [**1, 10**] for many explicit examples of conformal mappings. Both the theorem itself and the many explicit examples serve useful purposes in applied mathematics.

5.1. Non-Euclidean geometry. We next discuss the group G of conformal mappings from the unit disk **B** to itself and apply the ideas to non-Euclidean geometry. This group plays the same role in the hyperbolic geometry of **B** as the rigid motions do in the Euclidean plane $\mathbf{R}^2$.

The group G is *transitive*. In this context, transitivity means the following: given any two points w_1 and w_2 in **B**, we can find a conformal map of **B** mapping w_1 to w_2. To show that doing so is possible, we need only show that we can map 0 to an arbitrary point a. Then the inverse map takes a to 0. By composing a map taking w_1 to 0 with a map taking 0 to w_2, we get a map taking w_1 to w_2 as desired.

The map taking a to 0 has appeared already in Exercise 2.4. There we showed, for $|a| < 1$ and $|z| < 1$, that

$$(28) \qquad\qquad \left| \frac{z - a}{1 - \overline{a}z} \right| < 1.$$

Put $\phi_a(z) = \frac{a-z}{1-\bar{a}z}$. By (28), ϕ_a maps **B** to itself. Also $\phi_a(a) = 0$ and $\phi_a(0) = a$. The linear fractional transformation ϕ_a is invertible. One way to see the invertibility is to use the formula for the inverse of such a transformation given in Chapter 3. In this case, however, it is easier simply to compose ϕ_a with itself. A routine computation gives $\phi_a(\phi_a(z)) = z$, and hence ϕ_a is its own inverse. It follows that ϕ_a is a conformal map of **B** to itself. One can also check directly that the derivative ϕ_a does not vanish on **B**.

▶ **Exercise 8.32.** Verify that ϕ_a is its own inverse.

The group G of conformal mappings on **B** includes also the rotations $z \to e^{i\theta}z$.

Theorem 5.2. *Let f be a conformal map from **B** to itself. Then there is a point $a \in \mathbf{B}$ and a point $e^{i\theta} \in S^1$ such that*

$$f(z) = e^{i\theta}\phi_a(z).$$

Perhaps the most remarkable thing about G is its connection to non-Euclidean geometry. We define a geodesic in **B** to be a line or an arc of a circle that intersects the boundary circle at right angles. Each such geodesic is the image of the line segment $(-1, 1)$ under an element of G. Then all of Euclid's postulates for geometry hold except the *parallel postulate*. Given a point p not on a geodesic L, there is more than one geodesic through p and not intersecting (parallel to) L. This situation gives perhaps the most convincing example of a non-Euclidean geometry, called hyperbolic geometry. In the next exercise you are asked to show that there are infinitely many such geodesics.

▶ **Exercise 8.33.** Consider the circle defined by $x^2 + (y - b)^2 = r^2$. Choose b and r such that $b - r > 0$ and such that this circle intersects the unit circle at right angles. Then this circle defines a geodesic L in **B**. Show that there are infinitely many geodesics passing through 0 that do not intersect L. Hint: Draw a picture!

We list Euclid's postulates in Definition 5.2. See [**12**] for an excellent succinct treatment of this material. Euclid's fifth postulate is stated in terms of angles, but it is equivalent to the parallel postulate. We use the terms line, line segment, right angle, and congruent as usual in elementary geometry. In this language, the parallel postulate states the following: given a line L and a point p not on L, there is precisely one line through p and parallel to L.

Definition 5.2. Euclid's postulates follow:

- Each pair of distinct points can be joined by a unique line segment.
- Each line segment is a subset of exactly one line.
- Given a point p and a positive radius r, there is a unique circle of radius r centered at p.
- Any two right angles are congruent.
- Consider three lines l_1, l_2, and L. Suppose that l_1 and l_2 intersect L at distinct points. Suppose that the interior angles on one side of L add up to less than 180 degrees. Then l_1 and l_2 intersect on that side of L.

The interested reader should reinterpret these postulates in the hyperbolic geometry of **B** and verify that the first four hold. In this setting line becomes geodesic, line segment becomes arc of a geodesic, and so on. Be careful about the definition of a *hyperbolic circle*. A Euclidean circle C centered at the origin is a hyperbolic circle, but a Euclidean circle centered at another point is not. Given C, its image under ϕ_a will be a hyperbolic circle centered at a. Exercise 8.34 asks you to figure out what it is. Conversely each hyperbolic circle can be moved such that its center goes to the origin and its image is an ordinary circle. Recall that the group G of conformal mappings of the unit disk is the hyperbolic analogue of the group of rigid motions of the plane, and hence G is used to define the word *congruent*.

▶ **Exercise 8.34.** What is the image of a (Euclidean) circle in **B** about 0 under the map ϕ_a?

6. Quaternions

In Chapter 1 we saw that a complex number can be regarded as a point in the plane. In particular the ability to multiply complex numbers enables us to multiply vectors in the plane. Within **C** we can divide by any nonzero number, and hence we can divide by nonzero vectors in the plane. In the 1840s William Rowan Hamilton attempted to find a method for multiplying and dividing nonzero vectors in three-dimensional space. Eventually, by introducing a fourth dimension, he stumbled upon the *quaternions*. See [**25**] for fascinating discussion about Hamilton's discovery of the quaternions **H**. See [**5**] for some of their uses in geometry and see [**2**] to glimpse their role in physics. These last two references are quite advanced. In this section we describe the quaternions from a naive perspective, aiming primarily to make connections with simple things we have seen already about **C**.

First let us recall some simple facts about **C**. Consider the four complex numbers $1, i, -1, -i$. They form a group under multiplication. We can abbreviate the information by the partial multiplication table

$$
\begin{array}{c}
\begin{array}{cc} 1 & i \end{array} \\
\begin{array}{c} 1 \\ i \end{array}
\begin{pmatrix} 1 & i \\ i & -1 \end{pmatrix}.
\end{array}
$$

Our first definition of a quaternion is simply a point in real four-dimensional space $\mathbf{R}^4$. Thus $\mathbf{v} = (a, b, c, d)$, where a, b, c, d are real numbers. We write

$$\mathbf{v} = (a, b, c, d) = a\mathbf{1} + b\mathbf{i} + c\mathbf{j} + d\mathbf{k},$$

where **1** stands for $(1, 0, 0, 0)$, **i** stands for $(0, 1, 0, 0)$, and so on.

Let **H** denote $\mathbf{R}^4$ with the following operations of addition and multiplication. We add vectors as usual:

(29) $$(a, b, c, d) + (A, B, C, D) = (a + A, b + B, c + C, d + D).$$

We multiply vectors in a manner similar to how we multiply complex numbers, but things are more complicated. First we set

(30) $$\mathbf{i}^2 = \mathbf{j}^2 = \mathbf{k}^2 = \mathbf{ijk} = -\mathbf{1}.$$

If we multiply the relation $\mathbf{ijk} = -\mathbf{1}$ on the right by $\mathbf{k}$ and then divide by -1, we obtain $\mathbf{ij} = \mathbf{k}$. Similarly we obtain $\mathbf{jk} = \mathbf{i}$. We also get $\mathbf{ki} = \mathbf{j}$. If we perform these multiplications in the opposite order, however, then minus signs arise. For example $\mathbf{ik} = -\mathbf{j}$. Thus the commutative law for multiplication fails.

We can remember these rules via the table

$$(31) \qquad \begin{array}{c@{\quad}c} & \begin{array}{cccc} 1 & i & j & k \end{array} \\ \begin{array}{c} 1 \\ i \\ j \\ k \end{array} & \begin{pmatrix} 1 & i & j & k \\ i & -1 & k & -j \\ j & -k & -1 & i \\ k & j & -i & -1 \end{pmatrix} \end{array}.$$

Now we can define multiplication by an arbitrary pair of quaternions. We multiply (a, b, c, d) by (A, B, C, D) by writing

$$(32) \qquad (a\mathbf{1} + b\mathbf{i} + c\mathbf{j} + d\mathbf{k}) * (A\mathbf{1} + B\mathbf{i} + C\mathbf{j} + D\mathbf{k}),$$

expanding by the distributive law, and using (30). Doing so yields a complicated formula:

$$(33) \qquad (a, b, c, d) * (A, B, C, D) =$$

$$(aA - bB - cC - dD, aB + bA + cD - dC, aC - bD + cA + dB, aD + bC - cB + dA).$$

We can express (33) using matrices. The transformation taking (a, b, c, d) into (33) has matrix

$$(34) \qquad M = \begin{pmatrix} A & -B & -C & -D \\ B & A & D & -C \\ C & -D & A & B \\ D & C & -B & A \end{pmatrix}.$$

Notice that the four columns of M are orthogonal vectors in $\mathbf{R}^4$.

This multiplication law almost makes $\mathbf{R}^4$ into a field. The additive identity $\mathbf{0}$ is of course given by $\mathbf{0} = (0, 0, 0, 0)$, and the additive inverse is given by multiplying each component by -1. The multiplicative identity $\mathbf{1}$ is $(1, 0, 0, 0)$. All the properties of a field hold except the commutative law: in general

$$\mathbf{vw} \neq \mathbf{wv}.$$

We have noted already for example that $\mathbf{ij} = \mathbf{k}$, but $\mathbf{ji} = -\mathbf{k}$. Sometimes the quaternions are called a *skew* field or a *noncommutative* field; we must remember however that $\mathbf{H}$ is not a field because the commutative law fails. Below we will see one striking difference when we consider square roots of $-\mathbf{1}$.

Just as with complex numbers, the easiest path to multiplicative inverses involves conjugation. We will see that many of the formal properties of complex numbers, often after subtle adjustments, hold for the quaternions. For example, we define conjugation by

$$(35) \qquad (a\mathbf{1} + b\mathbf{i} + c\mathbf{j} + d\mathbf{k})^* = a\mathbf{1} - b\mathbf{i} - c\mathbf{j} - d\mathbf{k}.$$

We then have

$$(36) \qquad \mathbf{v}^*\mathbf{v} = ||v||^2 = a^2 + b^2 + c^2 + d^2.$$

We must be careful about the order of multiplication. One checks that

$$(37) \qquad\qquad (\mathbf{vw})^* = \mathbf{w}^*\mathbf{v}^*.$$

Suppose that $\mathbf{v} \neq \mathbf{0}$. Then $\mathbf{v}$ has a multiplicative inverse, namely $\frac{\mathbf{v}^*}{||\mathbf{v}||^2}$. The parallel with $\mathbf{C}$ is striking. Furthermore we even have

$$(38) \qquad\qquad ||\mathbf{vw}||^2 = ||\mathbf{v}||^2||\mathbf{w}||^2.$$

We saw in Chapter 1 that an element of a field can have at most two square roots. In particular, in $\mathbf{C}$ there are two square roots of -1. In the quaternions there are infinitely many square roots of -1. Thus eliminating commutivity of multiplication significantly changes things!

Proposition 6.1. *A quaternion $\mathbf{v}$ satisfies $\mathbf{v}^2 = -1$ if and only if $\mathbf{v} = b\mathbf{i} + c\mathbf{j} + d\mathbf{k}$ where $b^2 + c^2 + d^2 = 1$.*

Corollary 6.1. *In $\mathbf{H}$ there is a one-to-one correspondence between square roots of $-\mathbf{1}$ and points on the three-dimensional unit sphere.*

This corollary gives some insight into why quaternions are important in geometry. We refer to [**5**] and its references for more information.

▶ **Exercise 8.35.** Prove (33) given (30).

▶ **Exercise 8.36.** Verify (36), (37), and (38).

▶ **Exercise 8.37.** Find the determinant of M in (34) and interpret the result.

▶ **Exercise 8.38.** Prove Proposition 6.1.

▶ **Exercise 8.39.** Prove that the product of two numbers, each of which is a sum of squares of four integers, is also a sum of squares of four integers. Suggestion: Use (36) and (38).

▶ **Exercise 8.40.** The eight quaternions $\pm 1, \pm \mathbf{i}, \pm \mathbf{j}, \pm \mathbf{k}$ form a group under multiplication. Write out the multiplication table.

▶ **Exercise 8.41.** Express formula (33) for quaternionic multiplication in terms of cross products and dot products. Comment: First write $\mathbf{v}$ as a pair (a, v) where $v \in \mathbf{R}^3$. Do the same for $\mathbf{V} = (A, V)$. Then express $\mathbf{vV}$ using $v \times V$ and $v \cdot V$.

7. Higher-dimensional complex analysis

In this final section we ask what an analytic function of several complex variables might be. Three definitions are plausible, and again they turn out to be equivalent. We start with complex Euclidean space $\mathbf{C}^n$, consisting of n-tuples $z = (z_1, ..., z_n)$ of complex numbers. The norm $|z|$ is defined as follows; its square is given by

$$|z|^2 = \sum_{j=1}^{n} |z_j|^2.$$

Then $|z - w|$ denotes the distance between points z, w in $\mathbf{C}^n$.

Let Ω be an open set in $\mathbf{C}^n$, and let $f : \Omega \to \mathbf{C}$. We mimic Section 1 from Chapter 6 in giving possible definitions of *complex analytic function*. The first possible definition of complex analytic is that, near each point p in Ω, f is given by a convergent power series. The second possible definition is that f is continuously differentiable and $\frac{\partial f}{\partial \bar{z}_j} = 0$ for $j = 1, ..., n$. The third definition is that f is complex differentiable in each coordinate direction; thus f is complex analytic in each variable when the other variables are held fixed. These definitions turn out to be equivalent. The proofs are again based upon analogues of the Cauchy integral formula in several variables. A difficult result of Hartogs states that analyticity in each variable separately implies joint continuity, enabling one to verify the other definitions via the several variables analogue of the Cauchy integral formula. See [**13**] for details.

We want to illustrate one difference in the subjects. In one dimension, the region of convergence of a power series is a disk. We give some examples, all based on the geometric series, to show that such a result cannot hold in higher dimensions. See [**13**] for a precise result describing the possible domains of convergence of a power series in several complex variables.

Example 7.1. For $(z, w) \in \mathbf{C}^2$, put $f(z, w) = \frac{1}{1-zw}$. Then we have, for $|zw| < 1$, the geometric series

$$f(z, w) = \sum_{n=0}^{\infty} (zw)^n.$$

The region of convergence is the unbounded set defined by the inequality $|zw| < 1$.

Example 7.2. Replace z by z^a and w by w^b in the previous example and we get the region of convergence to be the region determined by $|z^a w^b| < 1$.

Example 7.3. Put $f(z, w) = \frac{1}{(1-z)(1-w)}$. The region of convergence of the series is then the set (called a polydisk) defined by the pair of inequalities $|z| < 1$ and $|w| < 1$.

One major difference between complex analysis in several variables and analysis in one variable is that the Riemann mapping theorem fails completely in dimension two or more. We say that domains are *inequivalent* if there is no bijective complex analytic mapping between them. In dimension two or higher, most (in a sense that can be made precise) domains topologically identical to a ball are inequivalent to a ball. Also polydisks and balls are inequivalent. It follows that the geometry of the boundary of a domain in higher dimensions matters. Hence complex analysis in several variables is even more geometric than it is in one dimension.

A second major difference arises from the zero-sets of complex analytic functions. We saw in Chapter 6 that the zero-set of a complex analytic function is a discrete set of points, unless the function vanishes identically (on a connected set). The zero-set of a complex analytic function f of n variables is an example of a complex analytic variety; if the function is not identically zero, then this variety has complex dimension $n - 1$. In particular, if it is not empty, the set of points where $\frac{1}{f}$ has a singularity is a variety of positive dimension. These ideas lead to the Hartogs extension theorem; we state a simplified version. Assume $n \geq 2$. Take a ball Ω, and remove a closed and bounded subset K such that what is left is connected.

Assume that f is complex analytic on the complement of K in Ω. Then there is a complex analytic function F, defined on all of Ω, that agrees with f where f is defined. We say that F extends f. In one dimension such a result is false. Remove a single point p from $\mathbf{C}$; the function $\frac{1}{z-p}$ is then analytic except at p but cannot be extended to be analytic in all of $\mathbf{C}$.

Despite the differences we have described, a unified theory of complex analysis exists. The ideas required in higher dimensions often borrow from the ideas used in one dimension, and on occasion the theory of several variables repays the debt by changing the way we think about one complex variable. The way we have regarded analytic functions as independent of $\overline{z}$ in this book provides a good example. See [6, 13, 14, 16] and their references for more information on complex analysis in several variables.

We close this section by discussing polarization. At various times in this book we considered polynomials in z and $\overline{z}$, and we treated these variables as independent. Doing so makes some readers feel uneasy. The procedure is justified by the following result, whose proof relies on functions of two complex variables. For simplicity we give a much less general statement than is possible. See Chapter 1 of [6] for more information.

Theorem 7.1. *Let H be complex analytic on $\mathbf{C}^2$. Assume for all $z \in \mathbf{C}$ that $H(z, \overline{z}) = 0$. Then $H(z, w) = 0$ for all z and w.*

Proof. By elementary facts in several variables, H has a convergent power series expansion:

$$(39) \qquad H(z, w) = \sum_{a,b=0}^{\infty} c_{ab} z^a w^b.$$

We will show that the coefficients c_{ab} in (39) all vanish. Put $z = |z|e^{i\theta}$ in (39). Since $H(z, \overline{z}) = 0$, we obtain

$$(40) \qquad 0 = \sum_{a,b=0}^{\infty} c_{ab} |z|^{a+b} e^{i(a-b)\theta}.$$

Set $a - b = n$ in (40) to get, where $n \in \mathbf{Z}$,

$$(41) \qquad 0 = \sum_{n,b} c_{(n+b)b} |z|^{n+2b} e^{in\theta}.$$

For each $k \in \mathbf{Z}$ we integrate $e^{-ik\theta}$ times the expression in (41) around a circle of radius R centered at 0. It follows for all n that

$$(42) \qquad \sum_{b} c_{(n+b)b} R^{2b} = 0.$$

The sum in (42) is a power series in R and is identically zero as a function of R; hence its coefficients all vanish. Thus $c_{ab} = 0$ for all a, b and H is identically zero. $\qquad\square$

The reader who knows Fourier series will note from (41) that we are expressing 0 as a Fourier series $\sum d_n e^{in\theta}$ and concluding that each d_n is zero.

Theorem 7.1 extends to functions complex analytic on $\mathbf{C}^n$. We give one example of polarization from linear algebra. A linear transformation $U : \mathbf{C}^n \to \mathbf{C}^n$ is *unitary* if it preserves inner products. Thus, for all z, w we have

(43) $$\langle Uz, Uw \rangle = \langle z, w \rangle.$$

A linear transformation U preserves distances if

(44) $$||Uz||^2 = ||z||^2.$$

If U is unitary, it obviously preserves distances; in fact the two conditions are equivalent by polarization. Suppose U preserves distances. We express (44) in terms of coordinates and replace $\overline{z_m}$ by $\overline{w_m}$. We obtain

$$\sum_{j,k,l,m} u_{jk} z_k \overline{u_{lm} w_m} = \sum_{j,l} z_j \overline{w_l}$$

and conclude that U is unitary.

Further reading

The books [**1, 17, 19, 23**] are all fantastic treatments of complex analysis in one variable. Each of them is masterfully written and covers far more than we do here. On the other hand we spend much more time on basic material and we provide many elementary examples and applications, thereby providing a solid introduction to any of these books. The book [**22**] is an eccentric yet brilliant treatment of complex analysis. The book [**10**] is a standard text on complex analysis and it includes more exercises and applications than the above books. The accessible book [**18**] includes many applications, many exercises, and even some color pictures of fractals. The elementary texts [**3, 4, 8, 20**] provide all the prerequisite algebra and analysis used here. The books [**6, 9**] are more specialized but accessible and each can be studied after reading this book. The book [**13**] on complex analysis in several variables is sophisticated, but its Chapter 1 provides a compelling treatment of one complex variable from the perspective needed for studying complex analysis in higher dimensions. Material on several complex variables also appears in [**6, 14, 16**].

Bibliography

[1] Lars V. Ahlfors, *Complex Analysis: An Introduction to the Theory of Analytic Functions of One Complex Variable, third edition*, McGraw-Hill Book Co., New York, 1978.

[2] Michael Atiyah, *Collected Works: Vol. 5, Gauge Theories*, Oxford University Press, New York, 1988.

[3] Robert G. Bartle and Donald Sherbert, *Introduction to Real Analysis*, John Wiley and Sons, New York, 1982.

[4] G. Birkhoff and S. MacLane, *A Survey of Modern Algebra*, MacMillan Co., Toronto, 1969.

[5] John H. Conway and Derek A. Smith, *On Quaternions and Octonions: Their Geometry, Arithmetic, and Symmetry*, AK Peters, 2003.

[6] John P. D'Angelo, *Several Complex Variables and the Geometry of Real Hypersurfaces*, CRC Press, Boca Raton, 1992.

[7] John P. D'Angelo, *Inequalities from Complex Analysis, Carus Mathematical Monographs 28*, Mathematical Association of America, Washington, DC, 2002.

[8] J. P. D'Angelo and D. B. West, *Mathematical Thinking: Problem Solving and Proofs, second edition*, Prentice-Hall, Upper Saddle River, NJ, 2000.

[9] Harold M. Edwards, *Galois Theory, Graduate Texts in Mathematics 101*, Springer-Verlag, New York, 1984.

[10] Steven D. Fisher, *Complex Variables, corrected reprint of the second (1990) edition*, Dover Publications Inc., Mineola, NY, 1999.

[11] T. Frankel, *The Geometry of Physics: An Introduction, revised edition*, Cambridge Univ. Press, Cambridge, UK, 1997.

[12] Timothy Gowers, *Mathematics: A Very Short Introduction*, Oxford University Press, Oxford, UK, 2002.

[13] Lars Hörmander, *An Introduction to Complex Analysis in Several Variables, third edition*, North-Holland Publishing Co., Amsterdam, 1990.

[14] Howard Jacobowitz, *Real hypersurfaces and complex analysis*, Notices Amer. Math. Soc. **42** (1995), 1480–1488.

[15] Mikhail I. Kadets and Vladimir M. Kadets, *Series in Banach Spaces: Conditional and Unconditional Convergence, translated from the Russian by Andrei Iacob, Operator Theory: Advances and Applications*, Birkhäuser, Basel, 1997.

[16] Steven G. Krantz, *Function Theory of Several Complex Variables, second edition*, Wadsworth and Brooks/Cole Advanced Books and Software, Pacific Grove, CA, 1992.

[17] N. Levinson and R. Redheffer, *Complex Variables*, McGraw-Hill, New York, 1970.

[18] J. Mathews and R. Howell, *Complex Analysis for Mathematics and Engineering*, Jones and Bartlett Publishers, Sudbury, 1996.

[19] R. Narasimhan and Y. Nievergelt, *Complex Analysis in One Variable, second edition*, Birkhäuser, Boston, 2000.

[20] K. Ross, *Elementary Real Analysis: The Theory of Calculus*, Springer, New York, 2000.

[21] M. Spivak, *Calculus on Manifolds: A Modern Approach to Classical Theorems of Advanced Calculus*, W. A. Benjamin, Inc., New York-Amsterdam, 1965.

[22] J. Stalker, *Complex Analysis: Fundamentals of the Classical Theory of Functions*, Birkhäuser, Boston, 1998.

[23] Elias M. Stein and Rami Shakarchi, *Complex Analysis, Princeton Lectures in Analysis II*, Princeton University Press, Princeton, 2003.

[24] J. Stewart, *Calculus: Early Transcendentals, third edition*, Brooks/Cole Publishing, Pacific Grove, 1995.

[25] B. L. van der Waerden, *Hamilton's discovery of quaternions*, Mathematics Magazine **49** (Nov. 1976), 227–234.

Index